Terramechanics

Terramechanics

Land Locomotion Mechanics

Tatsuro Muro

Department of Civil and Environmental Engineering
Faculty of Engineering
Ehime University
Matsuyama, Japan

and

Jonathan O'Brien

School of Civil and Environmental Engineering
University of New South Wales
Sydney, Australia

CRC Press
Taylor & Francis Group

A BALKEMA BOOK

CRC Press
Taylor & Francis Group
6000 Broken Sound Parkway NW, Suite 300
Boca Raton, FL 33487-2742

© 2000 by Taylor & Francis Group, LLC
CRC Press is an imprint of Taylor & Francis Group, an Informa business

No claim to original U.S. Government works

ISBN-13: 978-0-367-39460-8 (pbk)

Visit the Taylor & Francis Web site at
http://www.taylorandfrancis.com

and the CRC Press Web site at
http://www.crcpress.com

Contents

Preface

Terramechanics is a field of study that deals with the physical mechanics of land locomotion. It concerns itself with the interaction problems that occur between terrain and various kinds of mobile plant. This book seeks to explain the fundamental mechanics of the vehicle terrain interaction problem as it relates to the operation of many kinds of construction and agricultural vehicles. In particular, this text seeks to clarify matters that relate to the problem of flotation, trafficability and the mobility of wheeled and tracked vehicles running on soft terrains. Within terramechanics, it is clear that the ability of a construction or agricultural machine to do effective work in the field (i.e. its workability) depend to a very large degree on the physical properties of the terrain and primarily upon the strength and deformation characteristics of the soil, snow or other material. Because of this, a number of people, who are involved in the manufacture and use of working machines, need to understand the mechanical properties of ground materials like sandy or clayey soil and snow as encountered in the construction or agricultural fields. These people include design engineers – typically employed by the original equipment manufacturers, civil engineers – typically employed by the machinery users – and agricultural engineers who are commonly employed by client and land user groups. Through the use of several practical examples, the author aims to introduce terramechanics as a fundamental learning discipline and to establish methods for developing the workability and efficiency of various types of machinery operating across various kinds of soft terrain.

The author also seeks to build up a design method for a new, rational, process for working systems design that can be applied by civil, mechanical and agricultural engineers engaged in day to day field operations. Alternately, the process methodology may be employed by graduate and undergraduate students of terramechanics to develop sound principles of analysis and design within their study programs. In terms of historical background, the modern academic discipline of terramechanics and land loco-motion mechanics may be considered to have been 'invented' by M.G. Bekker over a period of 20 years in the 1950s and 60s. Bekker's ideas were expounded in three classic works entitled 'Theory of Land Locomotion' (1956), 'Off-the-road Locomotion' (1960) and 'Introduction to Terrain-vehicle Systems' (1969). These works are now considered to be the original 'bibles' in this field. Following on from Bekker, many early studies relating to the performance of wheeled and tracked vehicle systems took an essentially empirical approach to the subject. In recent times, however, academic studies in this field, and in associated fields such as soil and snow mechanics, have evolved into highly sophisticated domains involving both theoretical mechanics as well as experimental truths. As a consequence, modern land-locomotion mechanics may be considered to be approaching a full mathematical and mechanical science.

In relation to this newly developing science, a systematic approach to the layout of this text has been adopted. Also, in the body of this work, the importance of properly conducted experiments to determine the soil-machine system constants, the significance of the size effect, the problem of the interaction between wheel, track belt and terrain and the validity of the simulation analysis method is stressed. In Chapter 1, the mechanical properties e.g. the compressive and shear deformation characteristics of soil materials in soft terrain and snow materials in snow covered terrain are analysed. Also, several kinds of test method which are used to judge correctly the bearing capacity and the trafficability of wheeled and tracked vehicle are explained in detail. In Chapters 2 and 3, the land locomotion mechanics of wheeled vehicle systems composed of both rigid wheel and flexibly tired wheels are analysed. New simulation analytical methods are also introduced that use relations between soil deformation, amount of slippage at driving or braking state, the force and moment balances among driving or braking force, compaction resistance and effective driving or braking force to predict performance. Following this, several analytical and experimental examples are given.

In Chapter 4, some terrain-track system constants that may be used to properly evaluate the problem of interaction between track belt and terrain materials are developed from track plate loading and traction test results. Following this, the most important problems of determining the optimum shape of track plate to develop maximum tractive effort and the size effect of track plates on a soft terrain are analysed in detail from the perspective of several experimental examples.

In Chapters 5 and 6, the land locomotion mechanics of a tracked vehicle equipped with rigid and flexible track belts are analysed for the case of straight-line forward and turning motions on a soft, sloped, terrain. In relation to the traffic performances of both the rigid tracked vehicle and flexible tracked vehicle, an analytical method is used to obtain the relationship that exist between driving or braking force, compaction resistance, effective driving or braking force and slip ratio or skid, and contact pressure distribution. These relations are developed on the basis of the amounts of slippage and sinkage of track belt, energy equilibrium equation and the balance equation of the various forces acting on the tracked vehicle during driving or braking action. Following this, a new simulation analytical method is employed to calculate the amount of deflection of the flexible track belt and the track belt tension distribution during driving or braking action. Subsequent to this, several analytical examples of the traffic performances of a flexible tracked vehicle running on pavement road, decomposed weathered granite sandy terrain, soft silty loam terrain and snow covered terrain are presented and discussed.

In conclusion, it is observed that in recent years there has been a remarkable growth in interest in the potential use of robotized and un-manned working vehicles in the field of construction and agricultural machinery. The emerging academic field of terramechanics has great potential value here since it can give important suggestions to designers for development of the workability and the trafficability of potential robotized machinery intended for operation on soft earth and other forms of soft terrain.

To conclude this preface, it is the sincere hope of this author that readers will find this work to be useful. It is hoped that the book will give reliable guidance and information to students, mechanical engineering designers, robotic-machinery developers and to field engineers. In the writing of this book, the author has obviously sought to explain the basic theoretical mechanics as clearly as he can but nevertheless he is aware that there may be many deficiencies in layout and exposition. For improvement, reader suggestions and feedback and are earnestly sought. In the production of a book many people are involved.

The author wants to express his particular and most sincere thanks to Mr. H. Mori of the Department of business, to Mr. T. Ebihara and Mrs. Y. Kurosaki of the editorial department of Gihoudo Press, and to Mr. K. Kohno of Ehime University.

To my Australian collaborator, Jonathan O'Brien of the University of New South Wales, I give special thanks for his very high quality final English translation, for his technical editing and for his many suggestions and constructive criticisms. Further, I would specially like to thank him for the very large amount of work done in relation to the production of the camera-ready version of the book. However, the primary responsibility for the technical content of this work and for any errors or omissions therein remains that of the senior author.

Tatsuro Muro
Matsuyama, Japan

Chapter 1

Introduction

1.1 GENERAL

Over the last century, the mobile construction and agricultural equipment domain has risen from a position of virtual non-existence to one of major industrial significance. Nowadays, for example, everyone is familiar with heavy construction equipment that utilise either tracked or wheeled means for the development of powered mobility. The collective technical term used to refer to the mobility means used to propel and manoeuvre machines over varying terrains is 'running gear'. Wheels and tracks are the predominant types of running gear used in the mobile equipment industry but they are not the only ones. Legged machines and screw propulsion machines, for example, can be utilised – but they are quite rare.

Given the industrial importance of mobile machinery, a field of study that analyses the dynamic relationship between running gear and the operating terrain is of consequent major importance. The modern field of study that addresses this subject area is called 'Terramechanics'. The output of this field of study is qualitative and quantitative information that relates to such questions as the off-road trafficability of a vehicle, its travel capacity and the bearing capacity of a particular piece of ground. This type of information is necessary for the design of mobile machines and for their continuous performance improvement.

A central interest in terramechanics is that, relatively shallow, piece of ground that lies under a vehicle and which can become directly involved in a dynamic interplay with the equipment's running gear. The mechanical and deformation properties of this piece of terrain under both compressional and shear loading are defining in being able to predict the composite vehicle terrain behaviour. Many studies have been carried out to elucidate the complex relationship between the very many characteristics of the terrain and those of the vehicle. The design of the dimension and shape of tracks and wheels is very important when vehicles have to operate upon and across natural surfaces. This is so that vehicles when they are on weak terrain or snow do not suffer from excessive amount of sinkage when the bearing capacity of the ground underneath them is exceeded. Overcoming the sinkage problem requires a study of the phenomenon of vehicle 'flotation'. The study of this factor necessitates the study of the bearing capacity of the ground when the vehicle is both at rest and when it is in powered or unpowered motion.

Consideration of the vehicle in motion is required, since when the vehicle is in a self-propelling, driving or braking state, the wheel (or track) must effectively yield the shear resistance of the soil so as to generate the necessary thrust or drag. To ensure these generated products, a proper design of the grouser shape of the track and an appropriate selection of pressure distribution as well as the tread pattern and axle load of the tire is required. Further, it is necessary to study empirical results and field measurements if one wishes to maximize 'effective draw-bar pull' and/or to determine the forces required to overcome various motion resistances. These resistances include the slope resistance (which occurs when the vehicle is running up slopes) and compaction resistance (which is a consequence of the rut produced by static and slip sinkage).

The flotation, off-the-road trafficability and working capacity of a vehicle all depend principally on ground properties. In general, the ground property relating to the running of a vehicle is called 'Trafficability'.

On the other hand, the mobility of a piece of construction machinery depends on a host of vehicle parameters. These include factors that relate to engine power, weight of vehicle, spatial location of center of gravity, width and diameter of wheel, shape of tire tread, width and contact length of track, initial track belt tension, application point of effective draw-bar pull i.e. effective tractive effort or effective braking force, mean contact pressure, shape, pitch and height of grouser, diameter of front idler and rear sprocket, number of road rollers, suspension apparatus, type of connection and minimum clearance. In general, the running capability of a vehicle is called its 'mobility'.

The difference between the notions of 'trafficability' and 'mobility' is somewhat circular. 'Trafficability' may be thought of the ability (or property) of a section of terrain to support mobility. 'Mobility' is defined as the efficiency with which a particular vehicle can travel from one point to another across a section of terrain. To a degree, trafficability is a terrain characteristics centered concept whilst mobility is a task and vehicle-configuration centered concept.

Studies of both trafficability and mobility can provide a very useful background to the design and development of new construction machinery systems that aim to improve working capacity. These studies are most useful to those who wish to select a best or most suitable machine for a given terrain.

Now, let us turn to consideration of a number of experimental procedures that may be used to measure and to generally investigate the mechanical properties of the terrain surface of weak soil and of snow covered terrain. These experimental procedures can serve as the basis for a thorough study of the mobility characteristics of vehicles which may operate in either a driving or in a braking state. The methods also allow study of the turnability of tracked vehicles and the cornering properties of wheeled vehicles. Further, these approaches permit study of the travelling features of vehicles on slopes and allow study of the travelling mechanics of rigid or flexibly tracked vehicles or wheeled systems.

1.2. MECHANICS OF SOFT TERRAIN

In this section we will review some experimental field procedures that may be carried out to determine the soil constants for a particular terrain and which at the same time can help to define some of the physical properties, the shear resistance deformation characteristics, and the compression deformation characteristics of the soil. Knowledge of these soil properties is necessary to determine the trafficability viz. the thrust or the drag and the soil bearing capacity of tracked and wheeled vehicles which might be operating in a driving or braking state. The vehicles may be running on weak clayey soil or on loosely accumulated sandy soil.

In terms of available tests, use of the cone index to determine trafficability is of particular interest. The cone index will be discussed in more detail in later sections.

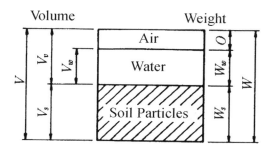

Figure 1.1. Idealised three-phase soil system.

1.2.1   Physical properties of soil

Natural soil typically consists of solid soil particles, liquid water and air in the void space. Figure 1.1 shows the typical structure of a moist natural soil in volumetric and gravimetric terms. The soil can also be thought of a being divided into two parts: a saturated soil part containing pore water and an unsaturated soil part.

The specific gravity of the solid soil particles may be measured according to the procedures of JIS A 1202 (say) by use of a pycnometer above 50 ml. The result is usually in the range from 2.65 to 2.80. If the volume of the soil particles is V_s (m³) their weight is W_s (kN), the void volume is V_v (m³), the volume of pore water is V_m (m³), the weight of pore water is W(kN), the total volume of the soil is V (m³) and the total weight of the soil is W (kN), then the unit weight γ is given by:

$$\gamma = \frac{W}{V} = \frac{W_s + W_w}{V} = G\gamma_w \quad (kN/m^3) \tag{1.1}$$

where G is the apparent specific gravity of the soil and γ_w is the unit weight of water. The void ratio e and the porosity of the soil n are non-dimensional and are defined as:

$$e = \frac{V_v}{V_s} \tag{1.2}$$

$$n = \frac{V_v}{V} \tag{1.3}$$

The water content w and the degree of saturation S_r are non-dimensional and are defined as:

$$w = \frac{W_w}{W_s} \times 100 \quad (\%) \tag{1.4}$$

$$S_r = \frac{V_w}{V_v} \times 100 \quad (\%) \tag{1.5}$$

In the case of saturated soil:

$$e = wG_s \tag{1.6}$$

For the case of an unsaturated soil, the following equation may be developed:

$$G = \frac{G_s(1+w)}{1+e} \tag{1.7}$$

Similarly formulae for the wet density ρ_t and the dry density ρ_d may be developed as follows:

$$\rho_t = \frac{W}{gV} \qquad (\text{g/cm}^3)$$ (1.8)

$$\rho_d = \frac{W_s}{gV} = \frac{\rho_t}{1 + w/100} \qquad (\text{g/cm}^3)$$ (1.9)

where g is the gravitational acceleration.

The relative density of a soil D_r can be expressed by the following equation where e_{max} is the maximum void ratio when the soil is loosely filled, e_{min} is the minimum void ratio when the soil is most compacted and e is the natural void ratio.

$$D_r = \frac{e_{max} - e}{e_{max} - e_{min}}$$ (1.10)

The value of the relative density for a loosely accumulated sandy soil lies in a range between 0 and 0.33.

The grain size analysis recommendations of national and international testing standards (such as JIS A 1204), indicate that the size of soil particles of diameter larger than 74μ can be determined by simple mechanical sieve analysis. Particles of diameters less than 74μ can be determined by sedimentation analysis using the hydrometer method.

Figure 1.2 shows an experimentally derived grain size distribution curve for a typical soil. The abscissa represents the diameter of soil particle on a logarithmic scale whilst the ordinate represents the percentages finer by weight. If D_{60}, D_{30} and D_{10} are diameters of the soil particles that have a percentage finer by weight of 60%, 30% and 10% respectively, then the coefficient of uniformity U_c and the coefficient of curvature U_c' are respectively:

$$U_c = \frac{D_{60}}{D_{10}}$$ (1.11)

$$U_c' = \frac{D_{30}^2}{D_{60} \times D_{10}}$$ (1.12)

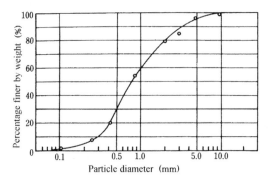

Figure 1.2. Particle size distribution curve.

In general, if the grain size distribution curve of a soil has a U_c is greater than 10 and if U_c' is in the range of $1 \sim 3$ and shows a S shape curve, then the soil is considered to have a good distribution [1].

Table 1.1 gives a classification of gravel, sand, silt and clay according to the Japanese unified soil classification system [2]. Alternatively, where one has determined the percentage of sand, silt and clay in a soil based on its grain size distribution curve, the name of that soil can be presented by use of a triangular soil classification system [3].

The angle of internal friction of a sandy soil depends mainly on its particle properties. However, it also depends on the shape of the individual soil particles and their particular surface roughnesses.

Figure 1.3 shows a typical sectional form of a soil particle. The slenderness ratio and the modified roundness of the sectional form [4] are defined as:

$$\text{Slenderness ratio} = \frac{a}{b} \tag{1.13}$$

$$\text{Modified roundness} = \frac{1}{2}\left(\frac{r_2 + r_4}{a} + \frac{r_1 + r_3}{b}\right) \tag{1.14}$$

where a and b are the apparent longer and shorter axes and $r_1 \sim r_4$ are the radii of curvature of the edge parts of the soil particle.

Table 1.1. Principal particle size scale [2].

1μm	5μm	74μm		0.42 mm	2.0 mm	5.0 mm	20 mm	75 mm	30 cm
Colloid			Fine sand	Coarse sand	Fine gravel	Medium gravel	Coarse gravel		
	Clay	Silt						Cobble	Boulder
			Sand			Gravel			
	Fine particle					Coarse particle			
								Rock materials	
				Soil materials					

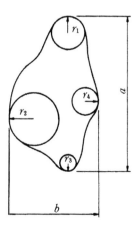

Figure 1.3. Measurement of slenderness ratio and modified roundness.

For groups of particles, these values should be expressed as an average. This is because (of necessity) the values need be determined from measurements taken from many individual particles. Measurement of group properties may be obtained by a sampling process wherein many representative particles are obtained by coating a microscope slide with a low cohesion glue. If the sticky slide is placed in the soil, thin layers of random particles will adhere to the slide. In scientific sampling work, slides need to be taken both in the horizontal and in the vertical plane of the soil.

In contrast to internal frictional values, the shear deformation characteristics of a cohesive soil depend largely on its water content. Thus, if we remove water from a cohesive soil in liquid state, it will transform into a more plastic state with an increase of cohesiveness. Further, with continued water content decrease it will become semi-solid and then solid. A measure of this physical metamorphism is called the 'Consistency' of the clay. Also, the limit of the water content is called the liquid limit w_L, the plastic limit w_p and the shrinkage limit w_s. These soil mechanics parameters are defined in many international and national standards. In Japan, the appropriate standards are JIA A 1205, 1206 and 1209.

The range of water contents over which a specimen exhibits plasticity can be expressed by a plastic index I_p, which is defined as:

$$I_p = w_L - w_p \tag{1.15}$$

Similarly, a consistency index I_c that shows the hardness and the viscosity of fine grained soil may be defined as:

$$I_c = \frac{w_L - w}{I_p} \tag{1.16}$$

When I_c approaches zero, the soil tends to be changed easily to a liquid state by disturbance from vehicular traffic. However, when I_c approaches one, the soil is a stable one because the soil is in an over-consolidated state with lower water content.

1.2.2 Compressive stress and deformation characteristics

A natural ground surface operated upon by a surface vehicle will typically be compressed and deformed as a result of the weight of the machinery passing into the soil through its wheels or tracks. Under rest conditions, the bearing characteristics of soil under the track of the construction machine, for example, can be calculated by the use of Terzaghi's limit bearing capacity equation [5,6]. This equation uses a coefficient of bearing capacity based upon the cohesion of the soil and its angle of internal friction. However, this equation can not be used to calculate the static amount of sinkage which is necessary to predict the trafficability of vehicles. Also, Boussinesq's method [7] for prediction of the soil stress in the ground based on the load on the semi-infinite elastic soil does not predict the static amount of sinkage under conditions where plastic deformation occurs – such as in weak terrains and in loosely accumulated sandy soils.

In contrast, the experimental plate loading test – as defined by JIS A 1215 – may be used as a direct field measurement of the coefficient of subgrade reaction of a section of road or base course. Using this method, direct measurement of the load sinkage curve is possible – provided that a standard steel circular plate of 30 cm diameter and thickness more than 25 mm is utilised.

In general, because the size of the loading plate has an effect on the amount of sinkage measured, it is necessary to develop some adjustment formula to allow correlation to the action of specifically sized wheels and tracks. More specifically, the following equation [8] can be employed to adjust for different contact areas of wheel and track. In the case of sandy soil, the amount of sinkage s (cm) of track of width B (m) may be calculated by use of the following formula.

$$s = s_{30} \left(\frac{2B}{b + 0.30} \right)^2$$ (1.17)

where s_{30} (cm) is the amount of sinkage of a standard $30 \times 30\,\text{cm}^2$ loading plate and the other part of the expression is a size adjustment factor. In the case of clayey soil, the amount of sinkage s (cm) for a track of width B (m) increases with an increase of B. It can be calculated for the same loading plate through use of the following formula [9].

$$s = s_{30} \frac{B}{0.30}$$ (1.18)

To relate the amount of sinkage s_0 (m) and the contact pressure $p = W/ab$ (kN/m^2) caused by a load W (kN) exerted on an $a \times b\,(a \geq b)\,(\text{m}^2)$ plate, Bekker [10] proposed the following equation (modifying the Bernstein-Goriatchkin's equation) which introduced the index n. Bekker suggested that the amount of sinkage increases with the width of the wheel or track according to the expression:

$$p = k s_0{}^n = \left(\frac{k_c}{b} + k_\phi \right) s_0{}^n$$ (1.19)

where n is a sinkage index (which varies depending on the type of soil) and k_c (kN/m^{n+1}) depends on the cohesion c, and k_ϕ (kN/m^{n+2}) depends on the angle of internal friction ϕ. The value of these sinkage coefficients can be determined by experiments using plates of two different b values [11].

The above equation – with size effect taken into consideration – reduces to Taylor's equation [12] when $n = 1$.

To obtain the values of the constants k and n, Wong [13] proposed the following F function (based on a least squares analysis method) which gives a weighting factor to the contact pressure p. This method takes as a basic assumption the idea that all observations have an equal reliability to error.

$$F = \sum p^2 (\ln p - \ln k - n \ln s_0)^2$$ (1.20)

More precise calculations of k and n can be achieved by solving the two expressions derived from the above equation when the partial differentials $\partial F/\partial k$ and $\partial F/\partial n$ are set to zero.

Taking a somewhat different approach, Reece [14] described the relationship between the amount of static sinkage s_0 (m) and the contact pressure p (kN/m^2) by the following equation.

$$p = (c k'_c + \gamma b k'_\phi) \left(\frac{s_0}{b} \right)^n$$ (1.21)

where c (kN/m^2) is the cohesion, γ (kN/m^3) is the unit weight, b (m) is the shorter length of the rectangular test-plate and k'_c and k'_ϕ are non-dimensional soil constants.

To complicate matters further, in the case when these equations are applied to actual tracks or wheels as loading devices the rectangular plate should be modified to correspond to the tire tread pattern or the grouser shape of the track. Furthermore, additional experiments must be executed to allow for inclined load effects under traction. Yet again, in cases where impact is applied to a relatively hard clayey soil or in cases where there exists punching action on a compound soil stratum, some different considerations apply.

To elucidate these matters, some experimental data for a model tracked-vehicle will be presented in Chapter 4. This work will discuss experimental determinations of the terrain-track constants from the plate loading test.

1.2.3 Shear stress and deformation characteristics

When a track or wheel is running free but is acted on by a positive driving or negative braking action and/or when the track or wheel acts upon a weak terrain, it experiences a phenomenon called 'Slip'. Slip involves relative motion of the running element and the soil and leads to the development of shear resistance due to the shear deformation of the soil. As a consequence, the soil exerts a thrust or drag corresponding to the amount of slippage necessary for the running of the vehicle. In addition, some degree of slip sinkage of the track or wheel will occur at the same time. This sinkage arises from a bulldozing type of action that develops as well to a volume change phenomenon that occur in the soil consequent to its shear deformation.

Let us now review some experimental methods and processes that can be used to determine the stress-strain curve and the volume change relation with strain for a soil. These two relations characterize the shear deformation relationship that applies to a particular soil.

(1) *Direct shear test*
Figure 1.4 shows a shear box apparatus comprised of two disconnected upper and lower boxes. The box can be filled with a soil sample at a certain density. If a static normal load P is applied to the specimen through a platen and the lower box is moved laterally a shearing action in the horizontal direction results. This action occurs over a shearing area A. The relationship between the volume change of the specimen (typically a small contraction or dilatation) and the shear deformation of the soil sample can be determined experimentally as can the relationship between the shear force T and the horizontal deformation of the lower box. This experimental method is formally referred to as the 'box shear test' [15].

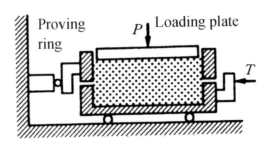

Figure 1.4. Direct shear box test apparatus.

In general, the force P is exerted through a gravity operated lever. The force T is typically measured by use of a proving ring equipped with an internal dial gauge. Measurement of the horizontal displacement of the lower box and the vertical displacement of the loading plate necessary for the measurement of the volume change of the soil samples can be carried out by the use of two dial gauges. The experiment is typically carried out at a shearing speed of between $0.05 \sim 2$ mm/min.

Figure 1.5 shows the general pattern of results that is achieved using this apparatus. In the diagram the relationship between shear resistance $\tau = (T/A)$ and horizontal displacement as well as the relationship between the volume change and the horizontal displacement is shown for conditions of constant normal stress $\sigma = (P/A)$. Typically, two types of experimental results are obtained for what one can refer to as type A and type B soils. In type A soils, the phenomenon is one where change of shear resistance τ and the contracting volume change tend towards a constant value or asymptote. This behaviour is typical of normally consolidated clays and loosely accumulated sandy soils. In contrast, the shear resistance τ of type B soils typically shows a marked peak at a certain horizontal displacement. Also, the expansive volume change occurs with a dilatancy phase following an initial contractional volume change. These hump type behaviours can be usually observed in overconsolidated clays and in compacted sandy soils. From these curves, a determination of the cohesion c and the angle of internal friction ϕ can be made through the use of Coulomb's failure criterion [16]. The process by which this is done is to repeat the shear test for a number of normal force values P, to record the peak shear resistance and then to plot the resulting data on a graph.

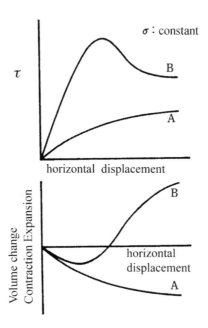

Figure 1.5. Relationship between shear strength, volume change and horizontal displacement.

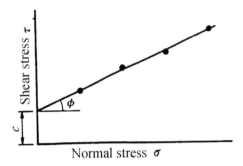

Figure 1.6. Determination of cohesive strength c and the angle of internal friction ϕ.

Figure 1.6 plots the peak shear resistance τ_{max} for a B type soil against the normal stress σ. The shear resistance τ value is that measured at a horizontal displacement of 8 mm (or 50% of the initial thickness of the soil sample) for an A type soil.

$$\tau = c + \sigma \tan \phi \tag{1.22}$$

The importance of this test can be appreciated from the fact that the bearing capacity of a terrain as well as the maximum thrust and the maximum drag of a tracked or wheeled vehicle system can be calculated using the cohesion c and the angle of internal friction ϕ.

One problem that arises here though, concerns the fact that the slip velocity of vehicles in practice may be greater than 1 m/s. This is an order of magnitude greater than that used in shear box test rate which is around 0.05 mm/min. This indicates that use of the latter shear rate to describe the actual shear deformation characteristics occurring under the wheels of real vehicle systems is clearly not appropriate.

In an attempt to overcome this loading rate difficulty, Kondo et al. [17] in 1986 developed a high speed shearing test for cohesive soils using a dynamic box shear test apparatus based on the principle of Hopkinson's bar. Their method, however, has not come into widespread use. A yet further deficiency of the standard box shear test, relates to the fact that the total shear displacement is much too small in comparison to the actual amount of slippage experienced with real tracked or wheeled vehicles. The arguments indicate why direct application of the relationships measured between shear resistance τ, volume change and horizontal displacement in the shear box to real world vehicle systems is not appropriate. Further, the shear box test does not include any of the bulldozing effects that typically occur in front of wheel systems. These methodological limitations makes effective prediction of the amount of slip sinkage that is associated with the slip of track or wheel impossible by use of this box shear method. To overcome some of these difficulties Chapter 4 will introduce an experimental procedure that is based upon a model tracked-vehicle. This experimental method is necessary to determine real value terrain-track system constants and is indispensable to the calculation of thrust and drag values as well as for calculation of the amount of slip sinkage that occurs under powered vehicles.

(2)   *Unconfined compression test*
The unconfined compression test is a test that is very commonly encountered in general materials testing and in soil mechanics. For example, JIS A 1216 describes the experimental procedures for carrying out an unconfined compression test in a standard manner. Using this

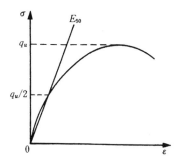

Figure 1.7. Relation between compressive stress σ and compressive strain ε.

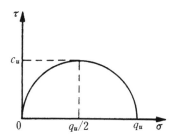

Figure 1.8. Mohr's failure circle.

procedure or similar ones (such are as prescribed by the ASTM or other testing authorities) the compressive stress–strain curve and the unconfined compressive strength q_u of a sample of a cohesive soil can be easily obtained. In the standard form of test, the relationship between compressive stress σ and compressive strain ε is measured at a compressive strain rate of 1%/min. The standard compression test specimen is a soil sample of cylindrical form. The specimen diameter is standardised at 35 mm and the height is set at 80 mm. A typical stress–strain curve for a test on a cohesive soil sample is shown in Figure 1.7.

This method is obviously not appropriate to the testing of sands and other granular solids. The stress developed in the stress–strain curve at the point where ε reaches 15% is deemed to be the unconfined compressive strength q_u. A value of a modulus of deformation ε_{50} can be developed based on the inclination of a line which links the origin and a point corresponding to the deemed compressive strength $q_u/2$ on the stress–strain curve. A Mohr's failure stress circle for this case is shown in Figure 1.8. In the case of a saturated clayey soil, $q_u/2$ equals the undrained shear strength of the soil c_u.

Further to this, it is an empirical observation from the field that, in general, when a track or wheel shears a saturated clayey soil at high speed, the undrained shear strength is constant and does not depend on the confining pressure. This result develops because shearing action is taking place under undrained conditions – as a consequence of pore water drainage not occurring.

As a consequence of this, the undrained shear strength c_u is very important in making assessments as to the trafficability of vehicles over given terrains.

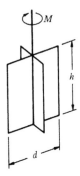

Figure 1.9. Vane shear test apparatus.

A sensitivity ratio S_t which expresses the sensitivity of a particular clayey soil to disturbance from passing vehicles can be developed using the following formula:

$$S_t = \frac{q_u}{q_{ur}} \tag{1.23}$$

where q_{ur} is the unconfined compressive strength of the undisturbed clayey soil and q_{ur} is the unconfined compressive strength when it has been remolded with a water content identical to that of the undisturbed soil.

The value S_t of a cohesive soil typically lies in the range $2 \sim 4$. A cohesive soil with $S_t > 4$ is called a sensitive clay. The degree of sensitivity of a clayey soil also plays an important role in assessments of the trafficability of a terrain to vehicles.

(3)   *Vane shear test*
The vane shear test method is the most normal procedure employed for in-situ measurement of very soft or weak cohesive terrains. Figure 1.9 shows the key features of the vane test apparatus. The vane itself is comprised of four blades which are rectangular plates of height h and width $d/2$ intersecting at right angles to each other. Shearing action in a clayey soil occurs when the vane is rotated continuously after its penetration into the ground. The relationship between the cohesion c and the maximum shearing torque M_{max} necessary for the vane to shear cylindrically a particular clayey soil in a cylindrical fashion can be expressed through use of the following formula – providing that the frictional resistance between the soil and the vane axle is ignored.

$$c = \frac{M_{max}}{\pi \left(hd^2/2 + d^3/6 \right)} \tag{1.24}$$

In this formula the blade diameter d is usually 5 cm and the shape is set as $h = 2d$.

Generally, the shear strength of a clayey soil shows anisotropy in different directions. Indeed, there are cases where the shear strengths in the horizontal and vertical directions are substantially different. Although the phenomenon of soil shearing by a track belt (such as with a tracked vehicle) typically occurs in a horizontal direction, the shear strength obtained in the vane shear test mainly takes place in the vertical direction. To determine the horizontal shear strength τ_H, Bjerrum's equation [18] can be applied. He found that the ratio of the horizontal shear strength τ_H to the vertical one τ_V is in inverse proportion to the plastic index I_p. In the general case, the shear strength τ_β in an arbitrary direction β to the horizontal plane can be calculated using Richardson's equation

$$\tau_\beta = \tau_H \tau_V \left(\tau_H^2 \sin^2\beta + \tau_V^2 \cos^2\beta \right)^{-\frac{1}{2}} \tag{1.25}$$

Shibata [20] pointed out that the ratio of anisotropy τ_V/τ_H for vane shear strength increases remarkably with decreases in the plasticity of the cohesive soil. He also suggested that, for cohesive soils of plasticity index of around 10, the ratio is almost equal to the coefficient of earth pressure at rest.

A further problem in relation to real situations occurs because the tractive characteristics of track-laying vehicles running on a very soft ground (i.e. the thrust or the amount of sinkage) are greatly influenced by the phenomenon of increase of shear strength with depth. Typically, ground shear strength develops linearly with depth. This problem can be analysed by using large deformation elasto-plastic FEM [21] (finite element analytical methods) in which the change of the undrained shear strength and the elastic modulus in the direction of depth is measured by means of the vane shear test. Moreover, in relation to the effect of shear rate on vane shear strength τ_V, Yonezu [22] and Umeda [23] carried out an experiment where the rotation speed of the vane was in the range of $\omega = 1.75 \times 10^{-3} \sim 1.62 \times 10^{-1}$ rad/s to $8.73 \times 10^{-4} \sim 8.73 \times 10^{-3}$ rad/s. They showed that the relation between τ_V and ω could be expressed as:

$$\tau_V = a \log \omega + b \tag{1.26}$$

In this formula, a is a coefficient which relates to the type of cohesive soil and b is a value which varies depending on depth. This type of research is extremely important in the field of terramechanics as it assists in the development of a clear understanding of the rate effect on the shear resistance of cohesive soils under the action of tracks or wheels. This topic, however, still requires much further investigation and research.

(4) *Bevameter test*
Figure 1.10 shows a Bevameter test apparatus in which the loading plate may take the form of the grouser shape of a track or of the tire tread pattern of a wheel. Through application of a normal load through the plate, the Bevameter can be used to measure the amount of sinkage and can be used to obtain the pressure sinkage curve which characterizes the compressive stress and deformation relationship of a terrain. Thence, by applying a torque under a constant normal load, the Bevameter can be used to measure the relationships that exist between shear resistance, normal stress and the amount of slippage. Alternately, it may be employed to determine the relationships that exist between the amount of slip sinkage, the normal stress and the amount of slippage which characterise the shear deformation of the terrain. This experimental apparatus was first developed by Söhne [24] who used it to obtain the pressure sinkage curve by employment of an hydraulic loading cylinder mounted on a vehicle.

Following this, Tanaka [25] revised the method through employment of a cone or a ring loading plate. He also developed an automatic recording device to record the normal pressure and the shear resistance relations. Bekker [26] developed a test vehicle equipped with plate load test apparatus which he used to obtain the pressure sinkage curve. The test vehicle was also equipped with a Bevameter test apparatus with a ring shear head. This was used to obtain the shear resistance deformation curve for a constant load. In 1981, Golob [27]

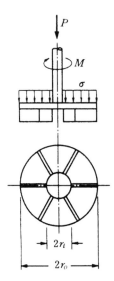

Figure 1.10. Bevameter test apparatus.

developed an apparatus which combined the plate loading test and the Bevameter test. This test apparatus can not only measure the pressure sinkage relations and the relationships that exist between shear resistance, normal stress and amount of slippage, but it can also measure the relationship between the amount of slippage, normal stress and amount of slippage.

The results of plate loading tests for plain plates or plates armed with grousers have already been mentioned in the Section 1.2.2. Let us now consider the case where a constant normal load P is applied to a ring head which is fitted with six vanes to simulate track grousers having certain grouser pitch and height ratio. In this case, the outer and inner diameters of the rings are defined as $2r_o$ and $2r_i$ respectively. A relationship between the shearing torque M and the displacement rotation j for a cohesive terrain can then be calculated by use of the following equation.

$$M = M_{max}\{1 - \exp(-aj)\}$$

$$j = \frac{r_i + r_o}{2}\alpha$$

(1.27)

where M_{max} is the maximum shear torque, a is the shape coefficient of the shear torque curve and α is the angle of rotation. So far as shear resistance τ is concerned, the shearing torque M may be calculated by the following equation:

$$M = \tau \int_{r_i}^{r_o} \left\{\pi(r + dr)^2 - \pi r^2\right\} rdr = \frac{2}{3}\pi\tau(r_o^3 - r_i^3)$$

(1.28)

This expression shows that the shearing torque M and the shear resistance τ have a proportional relation one to each other.

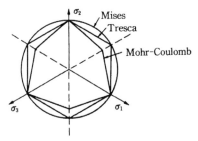

Figure 1.11. Failure curve.

If τ_{max} is the maximum shear resistance, then the relationship between the shear resistance τ and the rotational displacement of the Bevameter j for a cohesive terrain can be expressed as:

$$\tau = \tau_{max}\{1 - \exp(-aj)\} \tag{1.29}$$

Again, if the base area of the ring head is denoted as A, then the shear resistance–deformation curve which occurs under various applied normal stresses $\sigma = (P/A)$ can be determined.

In general, the associated amount of slip sinkage s_s – which is measured at the same time – can be expressed as a function of the normal stress σ and the amount of slippage of soil j_s. The relation can be expressed as:

$$s_s = c_0 \sigma^{c_1} j_s^{c_2} \tag{1.30}$$

where c_0, c_1 and c_2 are constants that relate to grouser pitch and height ratio and to the type of soil encountered. In this case $j_s = j$. The values for these particular constants can be obtained by micro-computer analysis of the results of the plate loading test and the ring shear test through application of the least squares method. An automatic data processing procedure for this technique has been developed by Wong [28].

(5) Triaxial compression test
With regard to the soil ingredients that comprise natural ground materials and within the general domain of soil mechanics, a number of constitutive equations [29] relating material stress and failure criteria [30] have been proposed.

Figure 1.11 shows the Von-Mises, Tresca and Mohr-Coulomb's failure criteria expressed in a coordinate system comprised of the three principal stresses σ_1, σ_2 and σ_3. The criteria themselves are as given in the following expressions.
[Tresca's criterion]

$$\tau_{max} = \frac{1}{2}(\sigma_{max} - \sigma_{min}) = \text{constant} \tag{1.31}$$

[Von-Mises's criterion]

$$\tau_{oct} = \frac{1}{3}\sqrt{(\sigma_1 - \sigma_2)^2 + (\sigma_2 - \sigma_3)^2 + (\sigma_3 - \sigma_1)^2} = \text{constant} \tag{1.32}$$

where τ_{oct} is the octahedral shear stress.

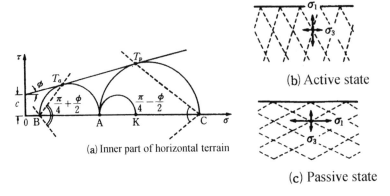

(b) Active state

(c) Passive state

(a) Inner part of horizontal terrain

Figure 1.12. Mohr-Coulomb failure criteria.

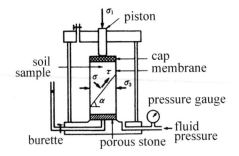

Figure 1.13. Triaxial compression test apparatus.

[Mohr-Coulomb's criterion]

$$\sigma_{max} - \sigma_{min} = 2c \cos \phi + (\sigma_{max} + \sigma_{min}) \sin \phi \tag{1.33}$$

$$\tau = c + \sigma \tan \phi \tag{1.34}$$

Most commonly, the well-known Mohr-Coulomb failure criterion is used to show the relationship between the shear resistance τ and the normal stress σ, as is shown in Figure 1.12. The initial stress condition in the ground can be defined by a Mohr's stress circle with a diameter of $\overline{AK}$. When a tracked or wheeled vehicle traverses the ground, the stress condition in the soil turns into an active or passive stress state.

These stress conditions can be represented by Mohr's stress circles that have diameters $\overline{AB}$ and $\overline{AC}$ respectively. The active or passive stress states which occur on the slip failure plane can be represented by T_a or T_p respectively in Figure 1.12. From the connecting line between T_a and T_p, (i.e. the failure envelope) the cohesion c and the angle of internal friction ϕ can be determined.

Figure 1.13 shows the essential features of the triaxial compression test apparatus. In the use of this apparatus, a sample of a clayey soil is formed into a cylindrical shape by means of a wire saw. The specimen is then sealed by covering its sides with a rubber membrane. The sample is then placed on a porous stone which is located at the base of the test apparatus and topped with a cap. The ends are then tightened towards each other.

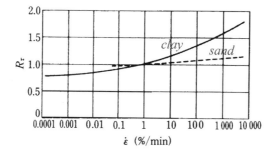

Figure 1.14. Relationship between strain rate and shear strength ratio R_t.

To determine the shear strength of a soil sample under actual drainage conditions, the sample is compressed with a vertical normal pressure σ_1 whilst the sample is confined laterally with a constant horizontal confining pressure σ_3 which is applied by hydraulic pressure. In this test arrangement, the relationship between the angle of slip failure plane α, the normal pressure σ and the shear stress τ can be expressed by the following equation.

$$\sigma = \frac{\sigma_1 + \sigma_3}{2} + \frac{\sigma_1 - \sigma_3}{2} \cos 2\alpha \tag{1.35}$$

$$\tau = \frac{\sigma_1 - \sigma_3}{2} \sin 2\alpha \tag{1.36}$$

Using these relations, the Mohr's stress circle can be determined as:

$$\left(\sigma - \frac{\sigma_1 + \sigma_3}{2} \right)^2 + \tau^2 = \left(\frac{\sigma_1 - \sigma_3}{2} \right)^2 \tag{1.37}$$

In the triaxial compression test, depending on the drainage conditions utilised the experimental method can be classed into three different test regimes – namely the unconsolidated undrained shear test (UU Test), the consolidated undrained shear test (CU Test) and the drainage shear test (D Test). In the field of terramechanics, the undrained shear test (UU Test) is commonly used because of the large strain rates encountered and the undrained condition typically encountered in the field. Although the standard triaxial compression test typically uses a quite fast strain rate of 1%/min, the required strain rate of the soil in terramechanics work is very much higher.

This is because there is a rapid shearing action which accompanies the dynamic loading of a ground by a track or a wheel. As a result, since the shear strength of a soil is largely dependent on the shear strain rate the actual strength of a material in the field will typically be quite a lot higher than that measured by the triaxial test.

Figure 1.14 illustrates the relationship [31] that might exist between strain rate and the ratio of shear strength to standard shear strength for both a clayey soil and a sandy soil sample – assuming a standard strain rate of 1%/min.

(6) Cone penetration test
Figure 1.15 shows a portable cone penetrometer apparatus to which is attached a cone of base area A and apex angle α. When the cone is forced into a ground with a speed of 1 cm/s the penetration force P (which varies with the penetration depth) can be measured

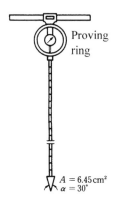

Figure 1.15. Cone penetrometer.

by use of a proving ring. This common piece of field instrumentation is called the 'cone penetrometer'. It is in world-wide use for assessing the trafficability of agricultural and construction terrains and machinery mobility. The penetration pressure acting on the base of the cone under constant penetration speed i.e. P/A is defined as the cone index CI. For cohesive soils, the following relation can be established between the cohesion c and the sensitivity ratio S_t [32].

$$CI = \frac{P}{A} = \frac{3\pi}{2}\left(1 + \frac{1}{\sqrt{S_t}}\right)c \qquad (1.38)$$

Although the apex angle α of test cones may take on values of $\pi/6$, $\pi/3$ and $\pi/2$ rad, it is an experimental finding that different values of this angle do not have a significant influence on the measured value of CI. However, because side friction can have a bearing on the measured result, systems have been developed that use a double tube.

Also, load cell based systems are available that can directly monitor, record and plot penetration force as a function of depth.

Hata [33] has analysed theoretically the relationship between the cone index CI and the cohesion c for both the initial stage and the steady state, taking into account the apex angle and the smoothness of the surface of cone. This analysis was carried out for the mechanism of the cone penetration into a cohesive terrain, considering it as a limit equilibrium problem. In a somewhat different vein, Yong et al. [34] studied the mechanism of cone penetration as a problem involving the mutual relationship between soil and wheel.

In general, where one has many passes of a piece of construction machinery in the same rut one gets remolding of a cohesive soil, with a consequent reduction of its shear strength. As a result the trafficability of the terrain will reduce.

The Japanese Association for Road Engineering's criteria for terrain trafficability are based on the cone index value that is necessary to allow a certain number of passes in the same rut. Table 1.2 gives a lower limit cone index value to permit the trafficability of various categories of construction machinery. In contrast, Cohron [35] has proposed a trafficability prediction method based on a comparison between the rating cone index RCI and the vehicle cone index VCI [36] that is necessary for one pass of a piece of construction machinery. The VCI value may be calculated from the mobility index MI [37] for a particular type

Table 1.2. Cone index necessary for trafficability of construction machinery.

Construction machine and situation	Cone index CI (kN/m^2)
Bulldozer running on very weak terrain	≥ 196
Bulldozer running on weak terrain	≥ 294
Middle size bulldozer	≥ 490
Large size bulldozer	≥ 686
Scrape-dozer running on weak terrain	≥ 392
Scrape dozer	≥ 588
Towed scraper	≥ 686
Motor scraper	≥ 906
Dump truck	≥ 1170

of construction machinery. The remolding index *RI*, which is the ratio of the rating cone index *RCI* of the soil after it has been subjected to the remolding produced by 50 vehicular passes and the cone index *CI* before remolding, is given by the following equation:

$$RI = \frac{RCI}{CI} \tag{1.39}$$

In general, for straight forward running motion across a flat terrain, if:

$RCI > VCI$ then vehicles can pass more than 50 times in the same rut
$RCI \geq VCI \times 0.75$ then vehicles can pass $1 \sim 2$ times in the same rut
$RCI \leq VCI \times 0.75$ then vehicles cannot pass.

1.3 MECHANICS OF SNOW COVERED TERRAIN

In this section we will present details of an in-situ experimental method that may be used to obtain snow characteristics and related snow physical properties. The method may also be used to obtain the snow's essential compression and shear deformation properties. These properties are necessary for calculation of the bearing capacity of a snow covered terrain or alternately of the thrust or drag of a tracked over-snow vehicle or wheeled snow remover during driving or braking action on a snow covered terrain. The covered terrain can consist of loosely accumulated newly fallen fresh snow or older sintered snow. Sintered snow is a snow which has been hardened by metamorphic action due to temperature or aging.

1.3.1 Physical properties of snow

Snow is composed of three phases – ice (solid), pore water (fluid) and void air (air). A typical composition is illustrated in Figure 1.16. Usually, snow can be classified into two types. The first type is wet snow at around $0\,°C$ when there is liquid pore water present. The second type is dry snow which exists at under freezing temperature. In this case there is no liquid pore water present. After fresh snow has fallen on the ground, the density of the snow will increase and become harder with time. This is due to the sublimation of the snow crystal particles and to bonds developing between the snow particles. This collective phenomenon

Volume   Mass

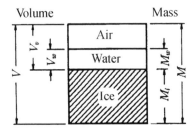

Figure 1.16. Idealised three-phase snow diagram.

is called snow metamorphism. During such a process a newly fallen snow of density of $0.05 \sim 0.10$ g/cm^3 can transform to a wet steady snow of density of $0.20 \sim 0.40$ g/cm^3 and thence to a granular snow of density $0.30 \sim 0.50$ g/cm^3.

Wet density ρ_w is defined as:

$$\rho_w = \frac{M}{V} \quad (\text{g/cm}^3) \tag{1.40}$$

where M is the mass of the sample and V is the volume of the snow specimen as sampled by an angular or cylindrical snow sampler and spatula.

Dry density ρ_i is defined as:

$$\rho_i = \frac{M_i}{V}\rho_w\left(\frac{100-w}{100}\right) \quad \text{g/cm}^3 \tag{1.41}$$

where w is the total weight water content expressed as a percentage. The water content itself is measured by use of a joint-type calorimeter which utilizes the latent heat of thawing ice. The final value may be calculated using the following equation:

$$w = \frac{M - M_i}{M} \times 100 = \frac{M_w}{M} \times 100 \quad (\%) \tag{1.42}$$

where M is the mass of the wet snow, M_i is the mass of the dry snow and M_w is the mass of the pore water.

Figure 1.17 shows the principle of operation of the widely used Yoshida joint-type calorimeter [38]. Firstly, hot water of mass M_A (g) and temperature T_1 (°C) is introduced into a heat insulating vessel A where it is kept warm. Next, a wet snow sample of 0 °C and mass of M_B (g) is taken from the snow covered area and put into another heat insulating vessel B. After this, the two vessels are connected and shaken to thoroughly mix the hot water and the snow sample. The temperature of the resulting mixture will become T_2 (°C). If the water equivalents of the vessel A and B are m_A (g) and m_B (g) respectively, the following identity can be established:

$$(T_1 - T_2)(M_A + m_A) = T_2(M_B + m_B) + JM_i \quad (\text{cal}) \tag{1.43}$$

where J is the latent heat of thawing ice ($= 79.6$ cal/g), and M_i (g) is the mass of ice in the wet snow sample M_B. Rearranging this equation we get:

$$M_i = \frac{1}{J}\{(T_1 - T_2)(M_A + m_A) - T_2(M + m_B)\} \tag{1.44}$$

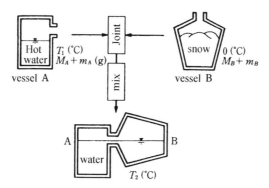

Figure 1.17. Measurement of water content of snow by use of a joint type calorimeter.

If one now substitutes M_i, calculated from the above equation, into Eq. (1.42), the water content w as a weight percentage can be obtained.

The degree of saturation S_r of a wet snow sample can be obtained from the ratio of volume of pore water V_w to the total volume V_v of the pore water and the pore air. The relation is as follows:

$$S_r = \frac{V_w}{V_v} \times 100 \tag{1.45}$$

The hardness of a snow covered terrain can be obtained by measuring the penetration depth of a cone or circular plate when a weight is dropped onto it from a pre-determined height. The push-type hardness measurement apparatus [39] can also measure the hardness of a snow covered terrain by measuring the diameter of the opening made by a cone of 20 cm diameter, height of 10 cm and apex angle of $\pi/2$ rad as it is spring-shocked by the apparatus.

A Canadian hardness meter, that is widely used in North America, measures the failure load of a snow covered terrain by use of a piece of equipment comprised of a circular plate attached to a cylinder that contains an in-built spring system. The plate is pushed into the snow and penetrates it.

A further type of system is the Rammsonde – which is a kind of cone penetrometer. In this method, the depth of penetration of a cone of $\pi/3$ rad apex angle is measured when a mass of 1 or 3 kg is dropped onto it from a prescribed height.

For the purpose of judging the trafficability of a snow remover or over-snow vehicle, Kinoshita's hardness meter [40] is widely used. This apparatus is shown in Figure 1.18. A mass m is dropped n times from a height h onto a circular plate and then the amount of sinkage d of the plate is measured. The plate has an area S and a mass M. The hardness of the snow covered ground H can then be computed by use of the following expression:

$$H = \frac{1}{S} \left\{ m \left(1 + \frac{nh}{d} \right) + M \right\} \tag{1.46}$$

This equation has been developed on the assumption that the total potential energies of the weight and the circular plate equate to the mechanical work done by the plate as it moves the snow against the penetration resistance. The hardness H can be viewed as being

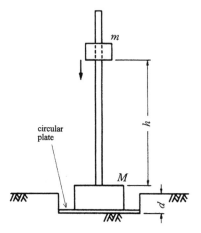

Figure 1.18. Kinoshita's hardness test apparatus.

the average penetration pressure under the circular plate. For naturally deposited snow and compacted snow, Kinoshita [41] concluded that the hardness of dry snow H (kPa) is proportional to the 4th power of the dry density ρ (g/cm^3). That is:

$$H = c\rho^4 \tag{1.47}$$

where the value of c is 100 for a naturally deposited snow and $400 \sim 600$ for a compacted snow.

Yoshida's experiment [42] explains why the hardness of wet snow decreases with an increase in water content as a percentage of total weight.

1.3.2 Compressive stress and deformation characteristics

Under the passage of snow removers or tracked over-snow vehicles, snow will become compressed. It will become even more compressed over time as the snow particles become bonded together by sintering. The relationship between the density of wet snow and the water content is very important in determining the engineering properties of compressed snow.

Figure 1.19 shows a series of results obtained where compaction of wet snow is obtained through the application of a constant compaction energy. Compaction in this case is obtained through use of special apparatus developed in conformance to JIS A 1210. From a study of this Figure, it can be concluded that the wet density ρ_w (g/cm^3) increases linearly with an increase in the total weight based, water content w (%) and can be expressed by the equation:

$$\rho_w = 9.98 \times 10^{-3}w + 0.54 \tag{1.48}$$

In contrast, the dry density ρ_i, can be seen in Figure 1.20, to not change with a change of, total weight-based water content w and it is evident that ρ_i remains at a constant value of $\rho_i = 0.57 \pm 0.05$ g/cm^3.

This encountered phenomenon is completely different from that found in relation to soil materials which have a maximum dry density at an optimum water content.

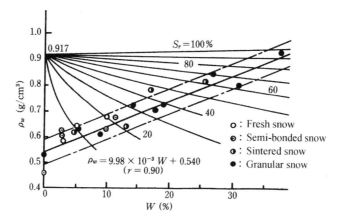

Figure 1.19. Relationship between wet density of wet snow compacted with constant energy and water content [43].

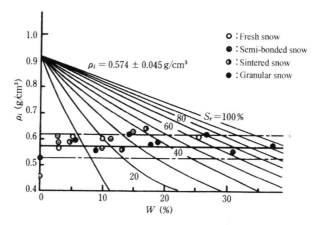

Figure 1.20. Relationship between wet density of wet snow compacted with constant energy and water content [43].

Next, let us consider the results that are obtained when a rectangular plate loading test [44] is performed on a snow covered terrain.

As shown in Figure 1.21, the main failure pattern resulting from a rectangular plate loading test on a dry fresh snow is that of the development of a compressed zone $\overline{bcfe}$ surrounded by two vertical slip lines $\overline{abc}$ and $\overline{def}$. As shown in Figure 1.22, the general relationship established between penetration pressure p and penetration depth X for six different kinds of dry fresh snow – identified as A_0, B_0, C_0, D_0, E_0 and F_0 having various initial densities at minus 13 °C – shows an elastic behaviour at the initial stage. After that, a plastic compression zone of snow develops under the rectangular plate and considerable micro-structural failures occur among the snow particles. These failures occur successively in front of the plastic compression zone. Subsequent to these two initial stages, we then get a phenomenon associated with plastic failure i.e. a saw-toothed hump. When the plastic

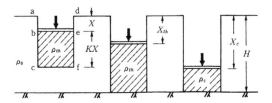

Figure 1.21. Compression and deformation process of fallen snow for plate loading.

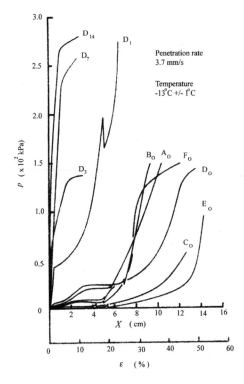

Figure 1.22. Relationship between penetration resistance, depth of penetration and compressive strain.

compression zone reaches the bottom of the snow box, the penetration pressure increases rapidly with increasing penetration depth.

In this case, the snow sample is assumed to be failed during the plate loading test since the strain controlled deformation rate of 3.7 mm/s is large enough as compared to a critical rate proposed by Kinoshita [45]. These phenomena have been observed in the plate loading test of Yoshida et al. [46] and have been observed in triaxial compression tests carried out at comparatively large deformation rate – cf. Kawada et al. [47].

In the initial elastic phase of the process, the relationship between the elastic modulus E (kPa) and the initial density ρ_0 (g/cm^3) of snow is given by the following relation:

$$E = 2642\rho_0^{2.826} \tag{1.49}$$

Table 1.3. Constant values from plate loading test for various snow samples.

Sample	ρ_d (g/cm^3)	E (kPa)	K	X_{th} (cm)	ε_{th} (%)	ρ_{th} (g/cm^3)	X_f (cm)	ε_f (%)	ρ_f (g/cm^3)
A_0	0.418 ± 0.017	153.4	3.800	5.72	20.9	0.512 ± 0.005	10.46	38.2	0.575
B_0	0.352 ± 0.022	213.7	3.000	6.86	25.0	0.469 ± 0.007	9.70	35.4	0.515
C_0	0.184 ± 0.017	43.8	5.353	4.32	15.8	0.229 ± 0.003	10.92	39.9	0.253
D_0	0.391 ± 0.014	104.1	4.425	5.08	18.5	0.467 ± 0.003	13.34	48.7	0.487
E_0	0.199 ± 0.021	16.4	5.750	5.33	19.5	0.262 ± 0.002	14.22	51.9	0.333
F_0	0.325 ± 0.010	76.7	4.564	4.95	18.1	0.390 ± 0.002	14.22	51.9	0.333
D_1	0.406 ± 0.001	2529.0	3.154	6.60	24.1	0.524 ± 0.003	–	–	–
D_3	0.397 ± 0.001	2058.0	2.708	–	–	0.491 ± 0.003	–	–	–
D_7	0.398 ± 0.002	5055.0	2.500	–	–	0.512 ± 0.007	–	–	–
D_{14}	0.398 ± 0.011	8505.0	2.200	–	–	0.522 ± 0.004	–	–	–

Because the thickness of the plastic compression zone is observed to be proportional to the penetration depth X, the coefficient of proportionality K can be regarded as a coefficient of propagation of plastic compression. Under these conditions the relationship between K and the initial density ρ_0 (g/cm^3) of snow can be expressed as follows:

$$K = 2.850 \rho_0^{-3.933} \tag{1.50}$$

When the plastic compression zone under the rectangular plate reaches the bottom of a snow covered ground of depth H, the penetration pressure p increases rapidly, takes a peak value at a penetration depth X_{th} and then decreases temporarily. The initial density of snow ρ_0 changes to a threshold density, as ρ_{th} proposed by Yong et al. [48] within the plastic compression zone and then it increases when the penetration depth X exceeds X_{th}.

The initial density finally develops to the density ρ_c of the polycrystallised ice when $X = X_{th}$. In this process, another compressive shear failure may occur due to the development of lateral plastic flow. If this happens, then the penetration pressure p will increase dramatically. Table 1.3 lists a number of constants for a variety of snow samples. The Table gives values for initial density ρ_0, elastic modulus E, the coefficient of propagation of plastic compression K, threshold penetration depth X_{th}, threshold strain $\varepsilon_{th} = X_{th}/H$ of the snow covered ground as well as threshold density ρ_{th}, final penetration depth of the rectangular plate X_f, final strain $\varepsilon = X_f/H$ of the snow covered terrain, and the final density ρ_f for a number of snow samples. Also given in the Table are the results of a series of rectangular plate loading tests for a number of sintered snow samples: D_1, D_3, D_7 and D_{14} (where the subscripted numbers denote the age of sintering) for a snow sample D conditioned under a temperature of minus 13 °C. As shown in the Table, penetration resistance under a constant penetration depth of the plate increases with increasing age of sintering. The fundamental failure pattern of the sintered snow can be explained in a similar way to that already mentioned in Figure 1.21 in that, it shows a remarkable elastic behaviour at the initial stage. After that, the failure pattern shows a complex behaviour as the snow changes its rheology and goes from an elasto-plastic to a rigid-plastic material as the penetration depth reaches some value.

1.3.3   Shear stress and deformation characteristics

To calculate the thrust or drag of a snow remover or tracked over-snow vehicle during driving or braking action, it is necessary to fully understand the shear deformation characteristics of snow covered terrains. Yong et al. [49] in 1978 performed a number of shear tests for a variety of snow materials. In their tests they used a shear box and established that the mechanism of shear in snow can be explained by consideration of the mutual adhesion and the mutual frictional forces that exist between the individual snow particles.

They also hypothesised that the Coulomb frictional relationship that connects normal stress and shear resistance could be used for artificially crushed snow particles but then the rule would not apply to sintered snows because the snow particles are bonded tightly due to sintering action.

Now, let us consider a test method for prediction of snow shear resistance based on the idea of use of a vane cone. Photo 1.1 shows a vane cone test apparatus as set up in a cold room of temperature of minus 13 °C. The test procedure involves making a sample by filling up a snow box. The vane or a vane cone as shown in Photo 1.2 is then pushed into the snow sample at constant penetration speed.

After measuring the relationship between the penetration force F and the penetration depth, another relationship between the torque T and the rotational deformation X can be determined. These factors can be measured as the vane or vane cone rotates and shears the snow sample. The test conditions are set at a peripheral speed of 1.89 mm/s and measurements are taken at a variety of depths.

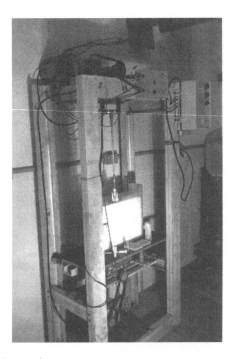

Photo 1.1. Vane cone test apparatus.

The merit of the vane cone test is that the relationship between the shear resistance τ and the shear deformation X can be measured directly under a constant normal stress σ.

In general, the shear resistance τ shows the shear deformation characteristic of a rigid plastic material in which the shear resistance goes immediately to a maximum value τ_{max} without any shear deformation and then decreases rapidly with increasing shear deformation X. The material follows the relation:

$$\tau = \tau_{max} \exp(-\alpha X) \tag{1.51}$$

where α is a constant value which is determined by the snow sample and the shear rate etc. If one assumes that the distribution of shear resistance τ acting on the vertical peripheral shear surface and the horizontal bottom shear surface of the vane cone of height of H and of radius r becomes uniform, then the normal stress σ and the shear resistance τ can be calculated from the penetration force F and the maximum torque T_{max}. This latter factor is measured in correspondence to the maximum shear resistance τ_{max} at $X = 0$ [50]. The calculated results can then be developed by use of the following equations.

$$\sigma = \frac{F}{\pi r^2} \tag{1.52}$$

$$\tau_{max} = \frac{T_{max}}{2\pi(r^2 H + r^3/3)} \tag{1.53}$$

As an example, Figure 1.23 shows the relationship between normal stress σ and shear resistance τ for a snow sample C [51].

Table 1.4 shows the relationships that exist between σ and τ for 6 kinds of fresh snow and for some samples of sintered snow. This data confirms that the Mohr-Coulomb failure criterion operates for the whole range of initial density, threshold density and final density of the fresh snow and for sintered snow samples. These results arise because the soil samples are tested in a state of normal compression as a consequence of the collapse of the structure of snow particles during the cone penetration.

Next let us consider the shear deformation characteristics of wet snow. Photo 1.3 shows a portable vane cone test apparatus. The shear deformation characteristics for a wet snow having the initial density of $\rho_0 = 0.228$ g/cm^3 were measured in the field using this apparatus. The experimental study showed the existence of an initial elasto-plastic behaviour

Photo 1.2. Vane and vane cone.

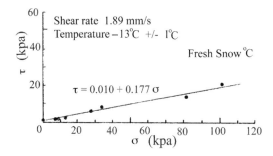

Figure 1.23. Relationship between shear stress and normal stress for the vane cone test [51].

Table 1.4. Relationship between shear stress and normal stress.

Sample	$\tau = f(\sigma)$ kPa
A_0	$\tau = 2.5 + 0.103\,\sigma$
B_0	$\tau = 3.4 + 0.277\,\sigma$
C_0	$\tau = 1.0 + 0.177\,\sigma$
D_0	$\tau = 4.3 + 0.418\,\sigma$
E_0	$\tau = 2.0 + 0.390\,\sigma$
F_0	$\tau = 2.1 + 0.548\,\sigma$
D_1	$\tau = 6.1 + 0.214\,\sigma$
D_3	$\tau = 8.8 + 0.245\,\sigma$
D_7	$\tau = 17.3 + 0.348\,\sigma$
D_{14}	$\tau = 13.6 + 0.222\,\sigma$

Photo 1.3. Portable vane cone apparatus.

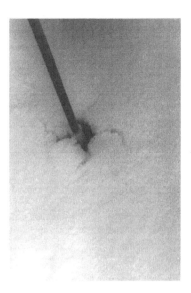

Photo 1.4. Slip line of snow vane cone test.

phase, in which the shear resistance took a maximum value at some shear deformation, and after that it decreased suddenly [52]. In this case, the slip line in the snow that developed during the vane cone shear test developed along a logarithmic spiral line. This spiral failure line is clearly visible in Photo 1.4. This shows clearly that snow materials are a kind of ϕ material.

The shear stress and deformation characteristics for the same wet snow were measured using the same portable shear test apparatus as shown in Figure 1.24 but to which was attached a ring of inner diameter of 0.5 cm and outer diameter of 3.5 cm and upon which was mounted a set of six vanes of height 1.0 cm.

Using this apparatus, the relationship between the shear resistance τ and the shear deformation y for the wet snow terrain – of threshold density $\rho_{th} = 0.275$ g/cm^3 and the critical density $\rho_C = 0.600$ g/cm^3 – could be experimentally explored. In this case, the study showed that the material behaves, phenomenologically, as an elasto-plastic material. Characteristic results for the tests are shown in Figure 1.25. In this case, the relationship between normal stress and the shear resistance also satisfies Coulomb's failure criterion. The angle of internal friction measured in this ring shear test is generally larger than that derived from the vane cone shear test for remolded snow samples.

1.4 SUMMARY

In this chapter we have introduced the idea of terramechanics as a study discipline and have located it at the interface between the two domains of machinery design/operation and ground behaviour.

We have also reviewed some traditional matters of soil mechanics as a precursor to applying them to machinery-terrain interaction problems. Also, we have analysed the mechanical

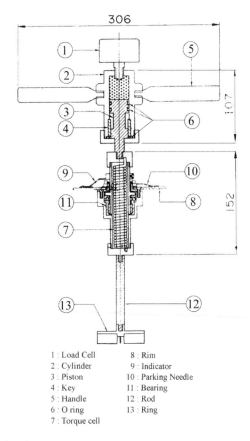

1 : Load Cell   8 : Rim
2 : Cylinder   9 : Indicator
3 : Piston   10 : Parking Needle
4 : Key   11 : Bearing
5 : Handle   12 : Rod
6 : O ring   13 : Ring
7 : Torque cell

Figure 1.24. Portable ring shear test apparatus.

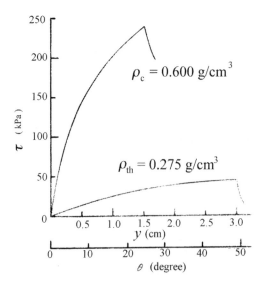

Figure 1.25. Relationship between shear stress and shear deformation.

properties of snow in some detail – since snow is a particular kind of terrain that is very important in some countries and in some latitudes.

Having studied this chapter, the reader should be able to describe the mechanical processes typically employed to, objectively and quantitatively, characterise specific terrains. The reader should also be able to appreciate and comment upon the degree to which the various terrain metrics can be used as inputs to engineering prediction models. These prediction models can be used for on-surface vehicle running behaviour prediction and/or for terrain trafficability studies.

REFERENCES

1. The Japanese Geotechnical Society (1983). *Soil Testing Methods*. (pp. 82–116). (In Japanese).
2. Yamaguchi, H. (1984). *Soil Mechanics*. (pp. 9–13). Gihoudou Press. (In Japanese).
3. Akai, K. (1986). *Soil Mechanics*. (pp. 3–26). Asakura Press. (In Japanese).
4. Oda, M., Enomoto, H. & Suzuki, T. (1971). Study on the Effect of Shape and Composition of Sandy Soil Particles on Soil Engineering Properties. *Tsuchi-to-Kiso*, Vol. 19, No. 2, 5–12. (In Japanese).
5. Toriumi, I. (1976). *Geotechnical Engineering*. (pp. 26–51). Morikita Press. (In Japanese).
6. Akai, K. (1964). *Bearing Capacity and Depression of Soil*. (pp. 5–34). Sankaido Press. (In Japanese).
7. Akai, K. (1976). *Advanced Soil Mechanics*. (pp. 201–234). Morikita Press. (In Japanese).
8. Terzaghi, K. & Peck, R.B. (1960). *Soil Mechanics in Engineering Practice*. (pp. 413–443). John Wiley & Sons and Charles E. Tuttle.
9. The Japanese Geotechnical Society (1982). *Handbook of Soil Engineering*. (pp. 484–486). (In Japanese).
10. Bekker, M.G. (1960). *Off-the-Road Engineering*. (pp. 25–40). The University of Michigan Press.
11. Sugiyama, N. (1982). *Several Problems between Construction Machinery and Soil*. (pp. 105–113). Kashima Press. (In Japanese).
12. Taylor, D.E. (1984). *Fundamentals of Soil Mechanics*. John Wiley & Sons.
13. Wong, J.Y. (1989). *Terramechanics and Off-road Vehicles*. (pp. 29–71). Elsevier.
14. Reece, A.R. (1964). Principles of Soil-Vehicle Mechanics. *Proc. Instn Mech. Engrs. 180, Part 2A(2)*, 45–67.
15. The Japanese Geotechnical Society (1983). *Soil Testing Methods*. (pp. 433–469). (In Japanese).
16. Leonards, G.A. (1962). *Foundation Engineering*. (pp. 176–226). McGraw-Hill and Kogakusha.
17. Kondo, H., Noda, Y. & Sugiyama, N. (1986). Trial Production of Dynamic Direct Shear. *Terramechanics*, 6, 98–104. (In Japanese).
18. Bjerrum, L. (1973). Problems of Soil Mechanics and Construction on Soft Clays and Structurally Unstable Soil (Collapsible, Expansive and Others). *Proc. of the 8th Int. Conf. on Soil Mechanics and Foundation Engineering*, Vol. 3, 111–159.
19. Richardson, A.M., Brand, E.W. & Memon, A. (1975). In-situ Determination of Anisotropy of a Soft Clay. *Proc. of the Conf., In-situ Measurement of Soil Properties. ASCE*, Vol. 1, 336–349.
20. Shibata, T. (1967). Study on Vane Shear Strength of Clay. *Proc. JSCE*, No. 138, 39–48. (In Japanese).
21. Muro, T. & Kawahara, S. (1986). Interaction Problem between Rigid Track and Super-Weak Marine Sediment. *Proc. JSCE, No. 376/III-6*. (In Japanese).
22. Yonezu, H. (1980). Experimental Consideration on Laboratory Vane Shear Test. *Tsuchi-to-kiso*, Vol. 28, No. 4, 39–46. (In Japanese).
23. Umeda, T., Ooshita, K., Suwa, S. & Ikemori, K. (1980). Investigation of Soft Ground using Vane of Variable Rotational Speed. *Proc. of the Symposium on Vane Test*. (pp. 91–98). The Japanese Geotechnical Society. (In Japanese).

24. Söhne, W. (1961). Wechselbeziehungen zwischen fahrzeuglaufwerk und boden beim fahren auf unbefestigter fahrbahn. *Grundlagen der Landtechnik, Heft 13*, 21–34.
25. Tanaka, T. (1965). Study on the Evaluation of Capability of Tractor on Paddy Field. *Journal of Agricultural Machinery*, 27, 3, 150–154. (In Japanese).
26. Bekker, M.G. (1969). *Introduction to Terrain-Vehicle Systems*. (pp. 3–37). The University of Michigan Press.
27. Golob, T.B. (1981). Development of a Terrain Strength Measuring System. *Journal of Terramechanics*, 18, 2, 109–118.
28. Wong, J.Y. (1980). Data Processing Methodology in the Characterization of the Mechanical Properties of Terrain. *Journal of Terramechanics*, 17, 1, 13–41.
29. Desai, C.S. & Siriwardane, J. (1984). *Constitutive Laws for Engineering Materials with Emphasis on Geologic Materials*. Prentice-Hall.
30. Mogami, T. (1969). *Soil Mechanics*. (pp. 492–506). Gihoudo Press. (In Japanese).
31. Skempton, A.W. & Bishop, A.W. (1954). Soils (pp. 417–482). Chapter 10 of *Building Materials*. Amsterdam: North Holland.
32. Muromachi, T. (1971). Experimental Study on the Application of Static Cone Penetrometer to the Investigation of Soft Ground. *Report on RailRoad Engineering*, No. 757. (In Japanese).
33. Hata, S. (1964). On the Relation between Cone Indices and Cutting Resistances of Cohesive Soils. *Proc. JSCE*, No. 110, 1–5. (In Japanese).
34. Yong, R.N., Chen, C.K. & Sylvestre-Williams, R. (1972). Study of the Mechanism of Cone Indentation and its Relation to Soil-Wheel Interaction. *Journal of Terramechanics*, Vol. 9, No. 1, 19–36.
35. Cohron, G.T. (1975). A New Trafficability Prediction System. *Proc. 5th Int. Conf., Vol. 2, ISTVS*, (pp. 401–425).
36. Sugiyama, N. (1982). *Several Problems between Construction Machinery and Soil*. (pp. 90–96). Kashima Press, (In Japanese).
37. Tire Society (1969). Considerations on Tire for Construction Machinery. *Komatsu Technical Report*, Vol. 15, No. 2, 1–54.
38. Yoshida, Z. (1959). Thermodynamometer Measuring Water Content of Wet Snow. *Institute of Low Temperature Science, Hokkaido University*, A-18, (pp. 19–28). (In Japanese).
39. Japanese Construction Machinery Association (1977). *New Handbook of Snow Prevention Engineering*. (pp. 11–58). Morikita Press. (In Japanese).
40. Kinoshita, S. (1960). Hardness of Snow Covered Terrain. *Institute of Low Temperature Science, Hokkaido University*, Physics, Vol. 19, (pp. 119–134). (In Japanese).
41. Kinoshita, S., Akitaya, E. & I. Tanuma (1970). Investigation of Snow and Ice on Road II. *Institute of Low Temperature Science, Hokkaido University*, Physics, Vol. 28, (pp. 311–323). (In Japanese).
42. Yoshida, Z. (1974). Snow and Construction Works (pp. 25–41). *Journal of the Japanese Society of Snow and Ice*. (In Japanese).
43. Muro, T. (1978). Compaction Properties of Wet Snow. *Journal of the Japanese Society of Snow and Ice*, Vol. 40, No. 3, (pp. 110–116). (In Japanese).
44. Muro, T. & Yong, R.N. (1980). Rectangular Plate Loading Test on Snow – Mobility of Tracked Oversnow Vehicle. *Journal of the Japanese Society of Snow and Ice*, Vol. 42, No. 1, 17–24. (In Japanese).
45. Kinoshita, S. (1967). Compression of Snow at Constant Speed. *Proc. of 1st Int. Conf. on Physics of Snow and Ice*, Vol. 1, No. 1, (pp. 911–927). The Institute of Low Temperature Science, Hokkaido University, Sapporo, Japan.
46. Yoshida, Z., Oura, H., Kuroiwa, D., Hujioka, T., Kojima, K., Aoi, S. & Kinoshita, S. (1956). Physical Studies on Deposited Snow, II. Mechanical properties (1), Vol. 9, (pp. 1–81). *Contributions from the Institute of Low Temperature Science, Hokkaido University*.
47. Kawada, K. & Hujioka, T. (1972). Snow Failure observed from Tri-Axial Compression Test. Physics, Vol. 30, (pp. 53–64), *Institute of Low Temperature Science, Hokkaido University*. (In Japanese).
48. Yong, R.N. & Fukue, M. (1977). Performance of Snow in Confined Compression. *Journal of Terramechanics*, Vol. 14, No. 2, 59–82.

49. Yong, R.N. & Fukue, M. (1978). Snow Mechanics – Machine Snow Interaction. *Proc. of the 2nd Int. Symp. on Snow Removal and Ice Control Research*, (pp. 9–13).
50. Muro, T. & Yong, R.N. (1980). Vane Cone Test of Snow-Mobility of Tracked Oversnow Vehicle. *Journal of the Japanese Society of Snow and Ice*, Vol. 42, No. 1, 25–32. (In Japanese).
51. Yong, R.N. & Muro, T. (1981). Plate Loading and Vane-Cone Measurements for Fresh and Sintered Snow. *Proc. of the 7th Int. Conf. of ISTVS, Calgary, Canada*, Vol. 3, (pp. 1093–1118).
52. Muro, T. (1984). Shallow Snow Performance of Tracked Vehicle. *Soils And Foundations*, Vol. 24, No. 1, 63–76.

EXERCISES

(1) The water content of a soil sampled in the field can be calculated from the difference between the weight of a wet and an oven dried soil sample. Suppose then, that the initial weight of a moist soil sample plus container was measured to be 58.2 g. Also, suppose that the total weight of the sample plus container – after oven drying for a period of 24 hours – was 50.6 grams. If the self weight of the container was 35.3 g, calculate the water content of the soil sample.

(2) Suppose that the grain size distribution curve for a particular soil sample is as shown in Figure 1.2, calculate the coefficient of uniformity U_c and the coefficient of curvature U'_c of the soil.

(3) Imagine that a plate loading test has been carried out to simulate a tracked machine's operation over a sandy terrain. The dimensions of the rectangular plate were: length 50 cm, width 20 cm and depth 5 cm. Assume also that results for the terrain system constants for Bekker's equation (1.19) which develops a relationship between the contact pressure p and the amount of sinkage s_0, have been obtained as $k_c = 18.2 \, \text{N/cm}^{n+1}$, $k_\phi = 5.2 \, \text{N/cm}^{n+2}$ and $n = 0.80$ respectively. Using this data, calculate the static amount of sinkage s_0 of a bulldozer of track length 1.5 m, track width 20 cm, and mean contact pressure 34.3 kPa.

(4) Suppose that a vane of diameter 5 cm and height of 10 cm is forced into a clayey terrain. After penetration the vane was rotated and a maximum torque value $M_{max} = 520 \, \text{N/cm}$ was obtained. Calculate the cohesion c of the clay, neglecting any adhesion between the rod of the vane and the clay.

(5) Given the following relationship between shear resistance τ and amount of slippage j, calculate the relationship between the modulus of deformation E and the constant value a.

$$\tau = \tau_{max} \{1 - \exp(-aj)\}$$

(6) Calculate the maximum amount of slip sinkage s_{max} when a tracked model machine of length 100 cm and width 20 cm loaded with a weight of 4410 N is pulled over a sandy terrain. The terrain-track system constants as given in Eq. (1.30) are $c_0 = 0.00476 \, \text{cm}^{2c_1-c_2+1}/\text{N}^{c_1}$, $c_1 = 2.07$, and $c_2 = 1.07$.

(7) To measure the weight percentage water content of a newly fallen wet snow, a joint type calorimeter as shown in Figure 1.17 was used and 2500 g of water at 85 °C was prepared and put into vessel A. Next, 300 g wet snow of 0 °C was sampled and put into vessel B. After joining both the vessels, the container was shaken until the temperature of the hot water assumed a constant value. If the temperature of the hot water decreased

to 55 °C due to melting of the snow sample, calculate the total weight based percentage water content of the wet snow. Assume a water equivalent of 400 g for vessel A and 300 g for vessel B.

(8) To measure the hardness of a snow covered terrain, Kinoshita's hardness testing machine is widely used. The mass of the falling weight is m, and the base area of the disc is S and its mass is M. Assuming that the total energy, when the falling weight falls n times from a constant height h equals the energy of penetration of the disc into the snow covered terrain, develop an equation to calculate the hardness H of the snow covered terrain.

(9) Assume that the total-weight based water content of a wet snow sample is measured as 20%. When this wet snow sample is compacted mechanically by a falling rammer at a constant compaction energy based on the criterion of JIS A 1210, calculate the compacted wet density ρ_w and the compacted dry density ρ_d of the snow sample.

(10) Suppose that a vane cone of radius of vane $r = 3$ cm and height $H = 12$ cm is penetrated fully into a snow covered terrain and that a maximum torque value of $T_{max} = 2.30$ kN/cm is measured under a penetration force of $F = 332$ N. Calculate the normal stress σ and the maximum shear resistance τ_{max} acting on the snow covered ground.

Chapter 2

Rigid Wheel Systems

For the compaction of subgrade materials, asphalt pavement materials and roller compacted concrete, several forms of compaction machines equipped with single steel drum rollers and tandem rollers are in everyday use in civil engineering – for example the macadam and steam rollers of old. The 'wheels' on these machines do not deform to any degree when they accept axle load, and as a consequence are referred to as 'rigid wheels'.

Al-Hussaini et al. [1] have analysed the distribution of contact pressures that will exist under a rigid wheel whilst it is running on a hard terrain. They analysed the distribution of normal stress and shear resistance using a procedure based on Boussinesq's theory. In Boussinesq's theory, the ground is assumed to be a semi-infinite elastic material. These researchers concluded that a network of stress lines corresponding to those of an equivalent normal and shear stress developed within the ground under a rigid wheel. In other research work, Ito et al. [2] measured the distribution of normal stress acting around a rigid wheel during driving action on a hard medium by use of a photo-elastic epoxy resin. They showed that the distribution of contact pressure could be divided into two parts; one is a compression zone which develops as a result of the thrust of wheel and the other is a bulldozing zone which produces increased land locomotion resistance with increasing vehicle sinkage.

Relative to the distribution of contact pressure of a rigid wheel running on a weak terrain, Wong et al. [3] showed that the application point of the maximum normal stress that develops on a sandy soil terrain could be expressed as a linear function of the slip ratio. They also showed that the distribution of shear resistance could be calculated from the slip ratio and that the normal stress distribution could be assumed to be symmetrical around the maximum-value application point .

Yong et al. [4–5] further analysed the distribution of contact pressure under a rolling rigid wheel by use of a theory of visco-plasticity that includes the strain rate behaviour of the soil. They also developed a method for predicting the continuous behaviour of a rigid wheel by use of the principle of energy conservation and equilibrium. They also attempted to analyse the stress response under a rigid running wheel using the finite element method (FEM). However, despite this research there are many unresolved problems which still need to be addressed in the future. These arise because it is still difficult to estimate the stress–strain relationship under a rigid wheel from the results of the plate loading test.

The trafficability of a rigid wheel running on a shallow snow covered terrain may be calculated from the total of the compaction energy – which is necessary to collapse the unremolded fresh snow under the roller – and the slippage energy which is required to give thrust to the wheel supplemented by the local melting of snow due to an excessive slip of wheel. Using this broad method, Harrison [6] developed a relationship between the effective tractive effort, amount of sinkage and the compaction resistance of a rigid wheel running over snow materials.

In the following sections of this Chapter, we will consider the various modes of behaviour of a rigid wheel running on weak ground during driving or braking action. We will then investigate the fundamental mechanics of land locomotion in relation to driving or braking force, thrust or drag, land locomotion resistance, effective driving or braking force, amount of sinkage and distribution of contact pressure.

2.1 AT REST

2.1.1 Bearing capacity of weak terrain

Where it is required to avoid an excessive amount of sinkage of a rigid wheel, such as when the wheel passes over a very weak terrain or a snowy ground, one needs to be able to predict the ground bearing capacity. The ground bearing capacity of a rigid wheel at rest Q_w can be calculated by use of the following formula for bearing capacity [7]. The formula assumes that the contact shape of the roller can be modelled as a rectangular plate of wheel width B and contact length L.

$$Q_w = BL \left\{ \left(1 + 0.3 \frac{B}{L} \right) cN_c + \left(0.5 - 0.1 \frac{B}{L} \right) B\gamma N_\gamma + \gamma D_f N_q \right\} \tag{2.1}$$

In this expression c and γ are the cohesion and unit weight of soil and γD_f is the surcharge corresponding to the amount of sinkage of wheel D_f. The factors N_c, N_γ and N_q are the Terzaghi coefficients of bearing capacity [8]. The factors may be determined from the cohesion and the angle of internal friction of the soil.

Relative to the stress conditions that prevail under a static rigid wheel, Terzaghi [9] predicted that the normal stress σ_Z in the vertical direction at a depth z under the corner of rectangular plate $B \times L$ to which is applied a uniformly distributed load q could be calculated as:

$$\sigma_Z = q \cdot I_\sigma$$

$$I_\sigma = \frac{1}{4\pi} \left\{ \frac{2mn(m^2 + n^2 + 1)^{\frac{1}{2}}}{m^2 + n^2 + m^2 n^2 + 1} \cdot \frac{m^2 + n^2 + 2}{m^2 + n^2 + 1} + \tan^{-1} \frac{2mn(m^2 + n^2 + 1)^{\frac{1}{2}}}{m^2 + n^2 - m^2 n^2 + 1} \right\} \tag{2.2}$$

where $m = B/z$ and $n = L/z$ are non-dimensional factors and I_σ is referred to as the factor of influence. The normal stress σ in the vertical direction at an arbitrary point in the interior of a rectangular plate may be calculated by a method that involves the sub-division of rectangles.

2.1.2 Contact pressure distribution and amount of sinkage

When a rigid wheel has an axle load W applied to it while it is standing at rest on a weak terrain, a reaction force develops that is a product of a symmetrical distribution of normal stress σ acting on the peripheral surface of the wheel in the normal direction and a shear resistance τ acting in the tangential direction of the wheel. Physically, the sign of the shear resistance τ should be reversed for the left and right hand sides of the wheels, such that the torque Q i.e. the integration of the shear resistance for the whole range becomes zero. A zero summation is required since no net torque on the wheel may be present.

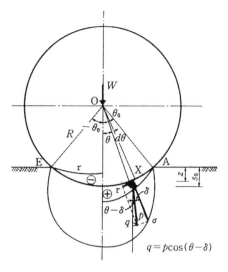

$$q = p\cos(\theta - \delta)$$

Figure 2.1. Distribution of ground reaction under a rigid wheel at rest.

If one considers the resultant stress p that develops between the normal stress σ and the shear resistance τ at an arbitrary point X along the peripheral surface of the contact part of wheel and terrain, the direction of application of the resultant stress p is must be inclined at an angle $\delta = \tan^{-1}(\tau/\sigma)$ relative to the normal direction. The general situation is as shown in Figure 2.1. The values of τ and δ depend on the amount of static slippage j between the rigid wheel and the terrain.

In this diagram, R is the radius of the rigid wheel, B is its width, $2\theta_0$ is the central angle $\angle AOE$, and θ is the central angle of an arbitrary point X on the contact part of the wheel. For the amount of sinkage s_0 of the bottom-dead-center of the wheel, the amount of sinkage z at the arbitrary point X can be calculated from geometry to be:

$$z = R(\cos\theta - \cos\theta_0) \tag{2.3}$$

Substituting $\theta = 0$ in the above equation, s_0 can be determined as,

$$s_0 = R(1 - \cos\theta_0) \tag{2.4}$$

When a rigid wheel penetrates into a ground under static load conditions, an interfacial shear resistance τ develops along the peripheral surface of the wheel to left and to the right hand sides of the center line. This interface tangential force develops by virtue of a relative motion (i.e. a slippage) that develops at the interface between the soil and the non-yielding penetrating wheel.

In the stationary wheel case, the amount of slippage j_0 can be assumed to develop as a function of the center angle θ as:

$$j_0 = R(\theta_0 - \sin\theta_0)\frac{\theta_0 - \theta}{\theta_0} \tag{2.5}$$

In general, the Mohr-Coulomb failure criterion for the maximum shear resistance τ_{max} to the normal stress can be used to analyse the interaction problem between a soil and a rigid wheel. In this situation:

$$\tau_{max} = c_a + \sigma \tan \phi \qquad (2.6)$$

Here, c_a and ϕ are parameters representing the adhesion and the angle of friction between the rigid wheel and the ground, respectively. The numerical value of these factors depends on the material that comprises the surface of the wheel, its surface roughness and the soil properties.

Most commonly, the shear resistance τ that develops along a rigid wheel running on a weak sandy or clayey terrain is smaller than τ_{max} and follows a relation as per Figure 2.2 where:

$$\tau = c_a' + \sigma \tan \phi' \qquad (2.7)$$

The parameter τ is the mobilized shear resistance which develops for a relative amount of slippage j between the rigid wheel and the terrain. It can be expressed as the product of a slippage function $f(j)$ times τ_{max} as follows:

$$\tau = \tau_{max} f(j) \qquad (2.8)$$

The slippage function $f(j)$ for a loose sandy soil and a weak clayey soil was given by Janosi-Hanamoto [10] as follows:

$$f(j) = 1 - \exp(-aj) \qquad (2.9)$$

In this expression the constant a is the value of the coefficient of deformation of the soil divided by the maximum shear strength τ_{max}.

For a hard compacted sandy soil and for a hard terrain, Bekker [11] and Kacigin [12] have proposed another function.

Working from these relations, the shear resistance $\tau(\theta)$ applied under the rigid wheel can be calculated as a function of the central angle θ as follows:

$$\tau(\theta) = \{(c_a + \sigma(\theta) \tan \phi)\} \{1 - \exp(-aj_0(\theta))\} \qquad (2.10)$$

Again, the vertical component $q(\theta)$ of the resultant stress $p(\theta)$ can be calculated as a function of the central angle θ from the plate loading test results through the use of Eq. (1.19). Note

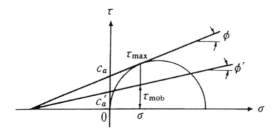

Figure 2.2. Relations between τ_{max}, τ_{mob}.

that allowance must be made here for the size effect of the contact length b. The result is the following expression:

$$p(\theta) = \frac{q(\theta)}{\cos(\theta - \delta)} = \frac{k_1 z^{n_1}}{\cos(\theta - \delta)} = \left(\frac{k_{c_1}}{b} + k_{\phi_1}\right) R^{n_1} \frac{(\cos\theta - \cos\theta_0)^{n_1}}{\cos(\theta - \delta)} \qquad (2.11)$$

Thence, the axle load W applied on the rigid wheel can be calculated by integration of the vertical component of the resultant stress $p(\theta)$ from $\theta = -\theta_0$ to $\theta = +\theta_0$ as follows:

$$W = BR \int_{-\theta_0}^{\theta_0} p \cos(\theta - \delta)\cos\theta \, d\theta \qquad (2.12)$$

Here, the central angle θ_0 i.e. the entry angle is required to be determined. First of all, the distribution of normal stress $\sigma(\theta) = p(\theta)\cos\delta$ and shear resistance $\tau(\theta) = p(\theta)\sin\delta$ have to be calculated for a given angle θ_0 and δ, and then the angle δ can be determined exactly from $\delta(\theta) = \tan^{-1}\{\tau(\theta)/\sigma(\theta)\}$.

These distribution of $\sigma(\theta)$ and $\tau(\theta)$ can be iteratively calculated until the real distribution of $\sigma(\theta)$, $\tau(\theta)$ and $\delta(\theta)$ is determined. Thereafter, the entry angle θ_0 can be determined by a repetition of calculation using the two division method until the axle load W is determined.

Then, the real entry angle θ_0, the real amount of sinkage s_0, and the distribution of normal stress σ and shear resistance τ can be determined.

As a further comment, it is noted that a horizontal force does not occur on a static rigid wheel, because the horizontal component of the resultant stress p is symmetrical on both the left and right hand sides.

2.2   AT DRIVING STATE

2.2.1   Amount of slippage

Figure 2.3 shows the distribution of slip velocity and amount of slippage respectively, for the contact part of a rigid wheel when it is in a driving state.

For a wheel of radius R, a moving speed V of the wheel in the direction of the terrain surface, and an angular velocity $\omega = -(d\theta/dt)$, the tangential slip velocity V_s at an arbitrary

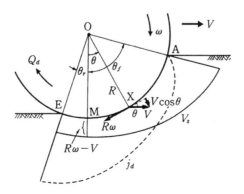

Figure 2.3. Distribution of slip velocity V_s and amount of slippage j_d during driving action ($R\omega > V$).

point X of central angle θ on the peripheral surface in the contact part of wheel to the terrain can be expressed as:

$$V_s = R\omega - V\cos\theta \tag{2.13}$$

where the angle θ is the angle measured from $\overline{OM}$ to the radius vector $\overline{OX}$. It is defined as positive for a counter clockwise direction. As $R\omega > V$ for the driving state, the slip ratio i_d is defined as follows:

$$i_d = 1 - \frac{V}{R\omega} \tag{2.14}$$

By substituting the above slip ratio i_d into Eq. (2.13), the slip velocity V_s at a point X can be expressed as:

$$V_s = R\omega\{1 - (1 - i_d)\cos\theta\} \tag{2.15}$$

After V_s takes a maximum value at the entry angle $\theta = \theta_f$, V_s decreases gradually to a minimum value $R\omega - V$ at the bottom-dead-center M and then increases slightly towards the point E at exit angle $\theta = -\theta_r$.

Integrating the slip velocity V_s from the beginning of contact $t = 0$ to an arbitrary time $t = t$, the amount of slippage j_d can be calculated as follows:

$$
\begin{aligned}
j_d &= R\int_0^t \omega\{1 - (1 - i_d)\cos\theta\}\,dt = R\int_\theta^{\theta_f}\{1 - (1 - i_d)\cos\theta\}\,d\theta \\
&= R\{(\theta_f - \theta) - (1 - i_d)(\sin\theta_f - \sin\theta)\}
\end{aligned} \tag{2.16}
$$

The amount of slippage j_d is positive for the whole range of the contact part $\overline{AE}$. It increases very rapidly from zero at point A at entry angle $\theta = \theta_f$. It takes a maximum value at point E at exit angle $\theta = -\theta_r$.

2.2.2 Soil deformation

Figure 2.4 shows the general soil deformation and the occurrence of slip lines when a rigid wheel is running on a sandy terrain during driving action. Typically, two slip lines emerge from the point N on the contact interface between the rigid wheel and the terrain. As developed in Terzaghi's theory of bearing capacity [13] the shape of the individual slip lines consists of a logarithmic spiral region and a straight line region.

When the slip line is in the region of the logarithmic spiral portion, the slip zone is referred to as the transient state zone. Similarly, the slip zone surrounding the straight slip line is called the passive state zone. The straight slip line usually crosses the terrain surface at an angle $\pi/4 - \phi/2$ for an angle of internal friction ϕ. As shown in Figure 2.4(a), the two slip lines s_1 and s_2 occur on both the left and right hand sides of point N. When the direction of one slip line s_1 is defined in the direction of the resultant velocity vector of V and $R\omega$, (which should occur at right angles to the moving radius $\overline{IN}$ for the instantaneous center N, the slip lines s_1 and s_2 cross at an angle $\pm(\pi/4 - \phi/2)$ to the major principal stress line σ_1 at the point N and at an angle $\pm(\pi/4 + \phi/2)$ to the minor principal stress line σ_3. The slipping soil mass NCA flows forward due to the down and forward resultant

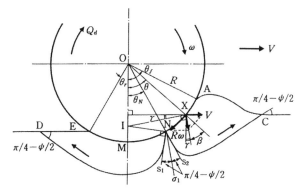

(a) Position of instantaneous center I

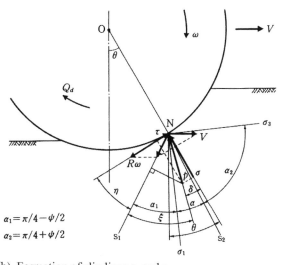

$\alpha_1 = \pi/4 - \psi/2$

$\alpha_2 = \pi/4 + \psi/2$

(b) Formation of slip lines s_1 and s_2

Figure 2.4. Soil deformation under a rigid wheel during driving action.

velocity vector, whilst the another slipping soil mass NDE flows backward due to the down and backward resultant velocity vector.

Within both slipping soil masses, there are a lot of conjugate slip lines s_1 and s_2 which introduce the occurrence of plastic flow within the soil. In addition, it should be noticed that the circumferential surface of the wheel is coincident directly with the slip line s_1 at the bottom-dead-center M.

The position of the point N and the direction of the slip lines depend on the slip ratio i_d which is calculated from the running speed V and the circumferential speed $R\omega$ of the wheel. In relation to this, Wong et al. [14] determined experimentally that the normal stress σ acting on a rigid wheel running on a sandy terrain takes a maximum value at point N. They also developed an experimentally based equation where the position ϕ_N of the point

N divided by the entry angle ϕ_f could be expressed as a linear function of slip ratio i_d as follows:

$$\phi_N/\phi_f = a + bi_d \qquad (2.17)$$

With reference to Figure 2.4(b), the angle η between the slip line s_1 and the tangential line at the point N can be calculated.

For the angle δ between the direction of resultant stress p and the normal line at the point N and for the angle ξ between the slip line s_1 and the resultant stress p:

$$\tan \eta = \frac{V \sin \theta}{R\omega - V \cos \theta} = \frac{(1 - i_d) \sin \theta}{1 - (1 - i_d) \cos \theta}$$

$$\xi = \frac{\pi}{2} - \delta - \eta$$

$$\therefore \frac{(1 - i_d) \sin \theta}{1 - (1 - i_d) \cos \theta} = \cot(\xi + \delta) \qquad (2.18)$$

As the angle ξ becomes the angle ϕ when the combination of the normal stress $p \cos \xi$ and the shear resistance $\pm p \sin \xi$ on the slip lines s_1, s_2 satisfies the failure condition of the soil, the angle θ_N of the point N can be derived theoretically using the above condition.

The resultant velocity vectors of all the soil particles on the contact part of the rigid wheel are always rotating around the instantaneous center I. The position of the center I is located between the axis O and the bottom-dead-center M for the driving state and the length $\overline{OI}$ can be given as follows, using the symbols of Figure 2.4.

$$\overline{OI} = R \cos \theta + R \sin \theta \tan \alpha = R \cos \theta + \frac{1}{\omega}(V - R\omega \cos \theta)$$

$$= \frac{V}{\omega} = R(1 - i_d) \qquad (2.19)$$

From the above equation, the length $\overline{OI}$ becomes the radius of wheel R when $V = R\omega$ i.e. $i_d = 0\%$, so the position of the instantaneous center I is coincident with the bottom-dead-center M. When the slip ratio i_d becomes 100%, $\overline{OI} = 0$, then the position of I is coincident with the axle point O. Therefore, the position of the center I depends on the slip ratio i_d, V and ω. This relationship between position of I and slip ratio i_d has been validated experimentally by Wong [15]. From general mathematics theory, it can be seen that the angle β between the resultant velocity vector and the normal line at the point X as shown in Figure 2.4(a) takes a minimum value when the point X is located at the position of $\angle OIX = \pi/2$.

Suppose that we now consider the locus of an arbitrary point on the peripheral surface of a rigid wheel when the wheel is running during driving action.

Figure 2.5 shows the rolling locus of an arbitrary point F on the peripheral surface of the rigid wheel when $V = R\omega$ i.e. the slip ratio i_d equals zero percentage. For a radius R, a moving velocity V and an angular velocity ω of the rigid wheel, the angle of rotation α of the point F is given as ωt for an arbitrary time t. During the rotation, the center of the wheel O moves to O' for the distance of $Vt = R\alpha$. The X, Y coordinates of the point F can

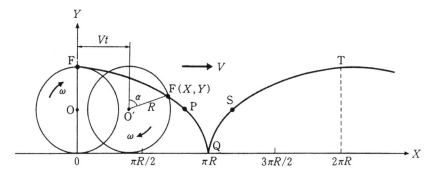

Figure 2.5. Cycloid curve for $V = R\omega$.

then be expressed as:

$$X = R(\alpha + \sin\alpha)$$
$$Y = R(1 + \cos\alpha) \tag{2.20}$$

These equations are those of a cycloid curve and the rolling locus follows a cycloid path, where the coordinates of the point P at $\alpha = \pi/2$ is $X = R(\pi/2 + 1)$, $Y = R$. Likewise the coordinate of the point Q at $\alpha = \pi$ is $X = \pi R$, $Y = 0$. Similarly, the coordinates of the point S at $\alpha = 3\pi/2$ is $X = R(3\pi/2 - 1)$, $Y = R$ and the coordinate of the point T of $\alpha = 2\pi$ is $X = 2\pi R$, $Y = 2R$.

The gradient of the tangent to the rolling locus dY/dX is coincident with the resultant velocity vector of the soil particle on the peripheral surface of a rigid wheel as follows:

$$\frac{dY}{dX} = -\frac{\sin\alpha}{1 + \cos\alpha} = -\tan\frac{\alpha}{2} \tag{2.21}$$

The gradient of the tangent to the rolling locus calculates to -1 at the point P, $-\infty$ at the point Q and then $+1$ and 0 at the point S and T, respectively.

Next, consider the problem of tracking the locus of an arbitrary point on the peripheral surface of a rigid wheel when the wheel is rolling at a slip ratio i_d during driving action. Figure 2.6 shows the rolling locus of an arbitrary point F on the peripheral surface of a rigid wheel running in the condition of $R\omega > V$. For an angle of rotation α at an arbitrary time t, α equals ωt and the center O of the wheel moves to O' for a distance $Vt = V\alpha/\omega$. Then, the X, Y coordinates of the point F can be expressed as:

$$X = \frac{V\alpha}{\omega} + R\sin\alpha = R\{\alpha(1 - i_d) + \sin\alpha\} \tag{2.22}$$
$$Y = R(1 + \cos\alpha) \tag{2.23}$$

These mathematical expressions are those of a trochoid curve. Therefore, the rolling locus of a point on a slipping wheel follows a trochoid shaped path, where the coordinates of a point P at $\alpha = \pi/2$ are $X = R\{(1 - i_d)\pi/2 + 1\}$ and $Y = R$. Similarly, the coordinates of the point Q at $\alpha = \pi$ are $X = \pi R(1 - i_d)$ and $Y = 0$.

Likewise, the coordinates of the point S at $\alpha = 3\pi/2$ are $X = R\{(1 - i_d)3\pi/2 - 1\}$, $Y = R$ and the coordinate of the point T at $\alpha = 2\pi$ are $X = 2\pi R(1 - i_d)$, $Y = 2R$.

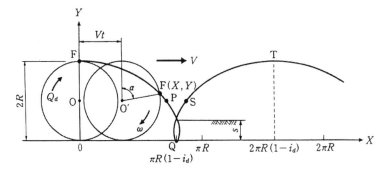

Figure 2.6. Trochoid curve during driving action ($R\omega > V$).

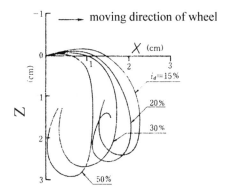

Figure 2.7. Moving locus of a soil particle under rigid wheel during driving action [17] where x represents horizontal displacement and z represents vertical displacement.

The gradient of the tangent to the rolling locus dY/dX is also coincident with the resultant velocity vector of the soil particle on the peripheral surface of a rigid wheel and may be expressed as follows:

$$\frac{dY}{dX} = -\frac{\sin \alpha}{1 - i_d + \cos \alpha} \tag{2.24}$$

The gradient of the tangent to the rolling locus is calculable as $-1/(1 - i_d)$ at the point P, 0 at the point Q, and $+1/(1 - i_d)$ and 0 at the points S and T respectively. From Figure 2.6, it can be seen that the rolling locus draws a small loop around the contact part of the rigid wheel and the terrain which is a function of the amount of sinkage s. Also it can be seen that a moving loss of $2\pi R i_d$ occurs due to the slippage during one revolution of the wheel. Yong et al. [16] used this rolling locus of a wheel as a boundary condition to an FEM study that they used to analyse the trafficability of a rigid wheel running on weak clayey terrain.

Yong et al. [17] also observed the moving locus of the soil particles in a soil bin when a rigid wheel was running during driving action. Figure 2.7 shows the results of observation of the moving locus of the soil particle at a depth of 1.27 cm for various values of slip ratio i_d

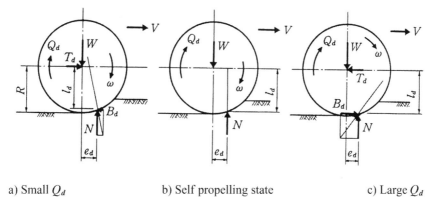

a) Small Q_d b) Self propelling state   c) Large Q_d

Figure 2.8. Ground reaction acting on rigid wheel during driving action.

when a rubber coated rigid wheel of weight 240 N was driven on a clayey soil. The moving locus of the soil particle show an elliptical path for each slip ratio.

Each soil particle moves forward and slightly upward during the progression of the wheel. It goes downward to a minimum position and then returns upward to a final position during the procession of the wheel. With increasing slip ratio i_d, the vertical distance of the final position of the soil particle i.e. the final amount of sinkage increases due to the increasing amount of slip sinkage and the horizontal distance of the final position tends to decrease.

2.2.3   Force balances

The traffic pattern of a rigid wheel when it is running during driving action at a constant moving speed V and a constant angular velocity ω can exist in three, quite distinct, modes or regimes depending on the magnitude of the driving torque $Q_d > 0$ that is applied to the wheel. The three operating regimes or modes of action are as shown in Figure 2.8.

Figure 2.8(a) shows that in mode 1 the effective driving force T_d occurs in the moving direction of the wheel when the driving torque Q_d is comparatively small. The effective driving force T_d balances with the horizontal ground reaction B_d which acts reversely to the moving direction of the wheel. The axle load W balances with the vertical ground reaction N. The direction of the resultant ground reaction, which is composed of N and B_d, does not go through the wheel axle, but deviates a little bit to the front-side of the wheel axle. The position of the point of application of the ground reaction on the peripheral surface of the rigid wheel has a horizontal position equal to the amount of eccentricity e_d and is located at a vertical distance from the axle of l_d.

The force balances in the horizontal and vertical directions, and the moment balance around the wheel axle can be established as follows:

$$W = N$$

$$T_d = B_d = \mu_d W$$

$$Q_d = N e_d - B_d l_d \qquad\qquad (2.25)$$

Then, the horizontal amount of eccentricity e_d can be calculated for a coefficient of driving resistance μ_d as,

$$e_d = \frac{Q_d}{W} + \mu_d l_d \tag{2.26}$$

Figure 2.8(b) shows a second mode of action where a rigid wheel is running in a self-propelling state. In this case, the effective driving force T_d and the horizontal ground reaction B_d decrease with an increase in driving torque Q_d. When T_d and B_d become zero, only the vertical ground reaction acts on the wheel. At this time, the force and moment balances and the horizontal amount of eccentricity e_d can be established as follows:

$$W = N$$

$$T_d = B_d = 0$$

$$Q_d = N e_d \tag{2.27}$$

$$e_d = \frac{Q_d}{W} \tag{2.28}$$

Figure 2.8(c) shows that, in a third mode of action, the effective driving force T_d occurs reversely to the moving direction of the wheel, when the driving torque Q_d increases by a very large degree. Sufficient effective tractive effort to be able to draw another wheel (say) will develop when the driving force can overcome the land locomotion resistance of the wheel. The direction of the horizontal ground reaction B_d is coincident with the moving direction of the wheel. The direction of the resultant ground reaction composed of N and B_d deviates to a large extent to the front-side of the wheel axle.

The force and moment balances and the horizontal amount of eccentricity e_d in this case can be established as follows:

$$W = N$$

$$T_d = B_d = \mu_d W$$

$$Q_d = N e_d + B_d l_d \tag{2.29}$$

$$e_d = \frac{Q_d}{W} - \mu_d l_d \tag{2.30}$$

For each of these three traffic patterns, the actual effective driving force T_d equals the horizontal ground reaction B_d and can be calculated as the difference between the horizontal component of the driving force $(Q_d/R)_h$ acting forward to the moving direction of the wheel and the land locomotion resistance L_{cd} acting backward to the moving direction. This relation can be expressed as follows:

$$T_d = \left(\frac{Q_d}{R}\right)_h - L_{cd} \tag{2.31}$$

In interpreting this expression, it can be seen firstly that the actual horizontal ground reaction B_d and its direction depends on the slip ratio i_d and secondly that B_d equals the difference between the driving force and the land locomotion resistance.

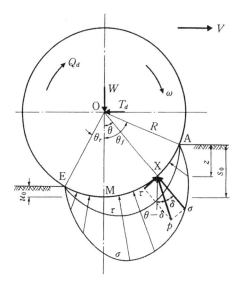

Figure 2.9. Distribution of contact pressure under rigid wheel during driving action.

2.2.4 Driving force

As shown in Figure 2.9, the distribution of the ground reaction applied to the peripheral surface of a rigid wheel during driving action can be expressed by the positive distribution of normal stress σ and shear resistance τ around the contact part $\widehat{AE}$, where A is the beginning point of contact to the terrain, and E is the ending point of departure from the terrain. The resultant stress p applies to the peripheral surface at an angle of $\delta = \tan^{-1}(\tau/\sigma)$ to the normal line. For an entry angle θ_f, exit angle θ_r and central angle θ between the vertical line $\overline{OM}$ and radius vector $\overline{OX}$ at an arbitrary point X on the peripheral surface of the rigid wheel, the amount of sinkage z for the central angle θ, the amount of sinkage s_0 at the bottom-dead-center M, and the amount of rebound u_0 after the pass of wheel can be expressed respectively as follows:

$$z = R(\cos\theta - \cos\theta_f) \tag{2.32}$$

$$s_0 = R(1 - \cos\theta_f) \tag{2.33}$$

$$u_0 = R(1 - \cos\theta_r) \tag{2.34}$$

The rolling locus of an arbitrary point X on the peripheral surface of the rigid wheel can be determined using Eq. (2.22) and Eq. (2.23).

The length of trajectory l can be obtained by integrating elemental sections of the rolling locus from $X = a$ to b as in the following formula:

$$l = \int_a^b \sqrt{1 + (dY/dX)^2}\,dX$$

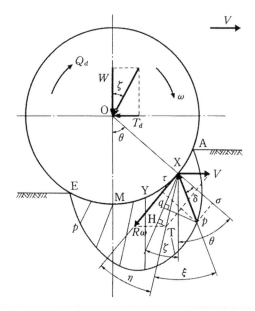

Figure 2.10. Component of a rolling locus in the direction of applied stress during driving action.

Substituting X, Y and dX/dY given in Eqs. (2.22), (2.23) and (2.24) into the above equation, the length of trajectory l can be expressed as $l(\alpha)$ which is a function of angle α, as follows:

$$l(\alpha) = \int_{\alpha_f}^{\alpha} \sqrt{1 + \left(-\frac{\sin\alpha}{1 - i_d + \cos\alpha}\right)^2} \cdot R(1 - i_d + \cos\alpha)\, d\alpha$$

Substituting the relation $\theta = \pi - \alpha$ and $\theta_f = \pi - \alpha_f$ into the above equation, the following further expression can be obtained:

$$l(\theta) = R \int_{\theta}^{\theta_f} \left\{(1 - i_d)^2 - 2(1 - i_d)\cos\theta + 1\right\}^{\frac{1}{2}} d\theta \tag{2.35}$$

As shown in Figure 2.10, the direction of the resultant force between the effective driving force T_d and the axle load W is given by the angle $\zeta = \tan^{-1}(T_d/W)$ to the vertical axis. In this case, the relationship between contact pressure $q(\theta)$ and soil deformation $d(\theta)$ in the direction of the resultant force agrees well with the plate loading test result previously given in Eq. (1.19). Here, $q(\theta)$ is the component of the resultant applied stress $p(\theta)$ to the direction of the angle ζ to vertical axis. $d(\theta)$ is the component of the rolling locus $l(\theta)$ to the same direction of the resultant force. That is, $d(\theta)$ is the component of the length of trajectory $l(\theta)$ in the direction of applied stress $q(\theta)$. $\overline{XT}$ is an elemental length of trajectory of $l(\theta)$ directed in the same direction as the resultant velocity vector of the vehicle velocity V and the circumferential speed $R\omega$. Then, $d(\theta)$ can be calculated as the integral of $\overline{XH}$ from $\theta = \theta$ to θ_f, which is the component of $\overline{XT}$ in the direction of the angle ζ to vertical axis, as follows:

$$d(\theta) = R \int_{\theta}^{\theta_f} \{(1 - i_d)^2 - 2(1 - i_d)\cos\theta + 1\}^{\frac{1}{2}}$$
$$\times \sin\left\{\theta + \zeta + \tan^{-1}\frac{(1 - i_d)\sin\theta}{1 - (1 - i_d)\cos\theta}\right\} d\theta \qquad (2.36)$$

The velocity vector V_p in the direction of the stress $q(\theta)$ at an arbitrary point X on the contact part of the peripheral surface of a rigid wheel and the terrain can be calculated as the component of the resultant velocity vector between V and $R\omega$, as follows:

$$V_p = \{(R\omega - V\cos\theta)^2 + (V\sin\theta)^2\}^{\frac{1}{2}}$$
$$\times \sin\left\{\theta + \zeta + \tan^{-1}\frac{(1 - i_d)\sin\theta}{1 - (1 - i_d)\cos\theta}\right\} \qquad (2.37)$$

Further, the plate loading and unloading test should be executed considering the 'size effect' of the contact length of wheel $b = R(\sin\theta_f + \sin\theta_r)$ and the 'velocity effect' of the loading speed V_p. The relationship between the stress component $q(\theta)$ of the resultant applied stress $p(\theta)$ and the soil deformation $d(\theta)$ can be determined as follows:

For $\theta_{max} \leq \theta \leq \theta_f$

$$q(\theta) = p(\theta)\cos(\zeta + \theta - \delta)$$
$$= k_1\xi\{d(\theta)\}^{n_1}$$

For $-\theta_r \leq \theta \leq \theta_{max}$

$$q(\theta) = p(\theta)\cos(\zeta + \theta - \delta)$$
$$= k_1\xi\{d(\theta_{max})\}^{n_1} - k_2\{d(\theta_{max}) - d(\theta)\}^{n_2} \qquad (2.38)$$

$$\xi = \frac{1 + \lambda V_p^{\kappa}}{1 + \lambda V_0^{\kappa}}$$

$$k_1 = \frac{k_{c1}}{b} + k_{\phi1} \quad \text{and} \quad k_2 = \frac{k_{c2}}{b} + k_{\phi2}$$

where θ_{max} is the central angle corresponding to the maximum stress $q_{max} = q(\theta_{max})$. The coefficients k_{c1}, $k_{\phi1}$ and k_{c2}, $k_{\phi2}$, and the indices n_1 and n_2 need to be determined from the quasi-static plate loading and unloading test for a loading speed V_0. The other coefficient λ and the index κ in the above coefficient of modification ξ for the previous Eq. (1.19) should be determined from a dynamic plate loading test at a plate loading speed of V_p.

Thence, the distribution of normal stress $\sigma(\theta)$ can be calculated as follows:

$$\sigma(\theta) = p(\theta)\cos\delta \qquad (2.39)$$

Also, the distribution of shear resistance $\tau(\theta)$ may be calculated by substituting the amount of slippage j_d given in Eq. (2.16) into the Janosi-Hanamoto's equation (2.10), as follows:

$$
\tau(\theta) = \{c_a + \sigma(\theta) \tan \phi\}
$$
$$
\times \left[1 - \exp\{-aR \left[(\theta_f - \theta) - (1 - i_d)(\sin \theta_f - \sin \theta)\right]\}\right] \tag{2.40}
$$

The Eq. (2.40) can be applied for a loose accumulated sandy soil or weak clayey terrain, but another equation [12] can be used for hard compacted sandy terrains and so on.

Following this, the angle $\delta(\theta)$ between the resultant applied stress $p(\theta)$ and the radial direction can be calculated as follows:

$$
\delta(\theta) = \tan^{-1} \left\{ \frac{\tau(\theta)}{\sigma(\theta)} \right\} \tag{2.41}
$$

Next, the axle load W, the driving torque Q_d and the apparent effective driving force T_{d0} can be coupled (using the force balance relations) to the contact pressure distribution $\sigma(\theta)$ and $\tau(\theta)$ as:

$$
W = BR \int_{-\theta_r}^{\theta_f} \{\sigma(\theta) \cos \theta + \tau(\theta) \sin \theta\} \, d\theta \tag{2.42}
$$

$$
Q_d = BR^2 \int_{-\theta_r}^{\theta_f} \tau(\theta) \, d\theta \tag{2.43}
$$

$$
T_{d0} = BR \int_{-\theta_r}^{\theta_f} \{\tau(\theta) \cos \theta - \sigma(\theta) \sin \theta\} \, d\theta \tag{2.44}
$$

and then the driving force can be calculated as the value of Q_d/R.

The apparent effective driving force T_{d0} as shown in the above equation can be expressed as a difference between thrust T_{hd} and compaction resistance R_{cd} in the following equation:

$$
T_{d0} = T_{hd} - R_{cd} \tag{2.45}
$$

where

$$
T_{hd} = BR \int_{-\theta_r}^{\theta_f} \tau(\theta) \cos \theta \, d\theta \tag{2.46}
$$

$$
R_{cd} = BR \int_{-\theta_r}^{\theta_f} \sigma(\theta) \sin \theta \, d\theta \tag{2.47}
$$

In this case, the amount of slip sinkage s_s is not considered. The thrust T_{hd} can be calculated as the integration of $\tau(\theta) \cos \theta$. It acts in the moving direction of the rigid wheel. On the other hand, the compaction resistance R_{cd} can be calculated as the integration of $\sigma(\theta) \sin \theta$. This force acts reversely to the moving direction of the wheel and manifests as the land locomotion resistance associated with the static amount of sinkage s_0.

As shown in the diagram, the point Y on the peripheral surface of the rigid wheel is determined when the direction of the resultant applied stress p becomes vertical. Hence it can be seen that the horizontal ground reaction acting on the section $\overline{AY}$ of the contact part to the terrain develops as the land locomotion resistance, whilst the horizontal ground reaction acting on the section $\overline{YE}$ develops as the thrust of the rigid wheel.

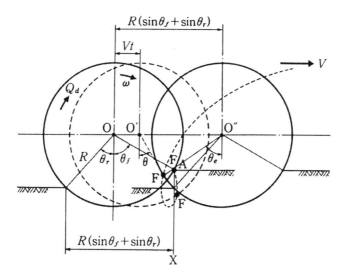

Figure 2.11. Rolling motion of a driven rigid wheel for a point X on the ground surface.

Additionally, the horizontal component of the driving force $(Q_d/R)_h$ is given as the summation of the actual effective driving force T_d and the total compaction resistance L_{cd} – which is the land locomotion resistance calculated considering the amount of slip sinkage.

2.2.5 Compaction resistance

In general, a considerable amount of slip sinkage will occur under a driven rigid wheel due to the existence of the dilatancy phenomena [18] associated with the shear action of soil at a peripheral interface.

From a plate traction test, the amount of slip sinkage s_s can be expressed as a function of contact pressure p and amount of slippage j_s as follows:

$$s_s = c_o p^{c_1} j_s^{c_2} \tag{2.48}$$

where the coefficient c_0 and the indices c_1, c_2 are the terrain-wheel system constants. These values have to be determined experimentally for a given steel plate and terrain.

Figure 2.11 shows the rolling motion of a rigid wheel during driving action as it moves around a point X on a terrain surface. To calculate the amount of slip sinkage s_s at the point X, it is necessary to determine the amount of slippage j_s at the point X.

When an arbitrary point F on the peripheral surface of the rigid wheel meets the point A on the terrain surface at the time $t = 0$, the central angle of radius vector $\overline{OF}$ to the vertical axis can be defined as the entry angle θ_f. Then, the central angle of the radial vector $\overline{OF'}$ to the vertical axis becomes θ at an arbitrary time t. The rotation angle ωt of the wheel can then be expressed as:

$$\omega t = \theta_f - \theta \tag{2.49}$$

Thence the moving distance $\overline{OO'}$ of the wheel during the time t can be calculated as:

$$\overline{OO'} = V_t = R\omega t(1 - i_d) = R(1 - i_d)(\theta_f - \theta) \tag{2.50}$$

The transit time t_e, as the rigid wheel passes over the point X, may be taken as the time when the distance $\overline{OO''}$ reaches the contact length of wheel $\overline{OO''} = R(\sin\theta_f + \sin\theta_r)$. The central angle of the radius vector $\overline{O''F}$ then becomes θ_e where:

$$\theta_e = \theta_f - \frac{\sin\theta_f + \sin\theta_r}{1 - i_d} \tag{2.51}$$

Substituting the above angle $\theta = \theta_e$ into Eq. (2.49) we get:

$$t_e = \frac{\sin\theta_f + \sin\theta_r}{\omega(1 - i_d)} \tag{2.52}$$

If we now recall, from the previous Section 2.2.2, that the amount of slippage of a rigid wheel during one revolution is $2\pi Ri_d$, then, the amount of slippage of soil j_s at the point X may be calculated as the ratio of t_e to $2\pi/\omega$ as follows:

$$j_s = 2\pi Ri_d \cdot \frac{R(\sin\theta_f + \sin\theta_r)}{2\pi R(1 - i_d)} = R(\sin\theta_f + \sin\theta_r)\frac{i_d}{1 - i_d} \tag{2.53}$$

If we now take a small interval of central angle θ_n and a small interval of amount of slippage j_s/N for a very small time interval t_e/N, then the vertical component of the resultant applied stress p can be calculated as $p(\theta_n)\cos(\theta_n - \delta_n)$ for the nth interval of the center angle θ_n. Then, by substituting these values into Eq. (2.48), the total amount of slip sinkage s_s can be calculated as the sum of the elemental amounts of slip sinkage as follows:

$$s_s = c_0 \sum_{n=1}^{N} \{p(\theta_n)\cos(\theta_n - \delta_n)\}^{c_1} \left\{ \left(\frac{n}{N}j_s\right)^{c_2} - \left(\frac{n-1}{N}j_s\right)^{c_2} \right\} \tag{2.54}$$

where

$$\theta_n = \frac{n}{N}(\theta_f + \theta_r)$$

Consequent to this calculation, a rut depth i.e. the total amount of sinkage of a rigid wheel s can be calculated from the static amount of sinkage s_0 given in Eq. (2.33). Also, the amount of rebound u_0 given in Eq. (2.34) and the above mentioned amount of slip sinkage s_s can be determined as follows:

$$s = s_0 - u_0 + s_s \tag{2.55}$$

Next, the total compaction resistance L_{cd} can be calculated if we make the assumption that the product of the compaction resistance L_{cd} applied in front of the rigid wheel and the moving distance $2\pi R(1 - i_d)$ during one revolution of the wheel can be equated to the rut making work which in turn may be calculated as the integration of the contact pressure p acting on a plate of length $2\pi R(1 - i_d)$ and width B from an initial depth $z = 0$ to a total amount of sinkage $z = s$,

$$2\pi R(1 - i_d)L_{cd} = 2\pi R(1 - i_d)B \int_0^s p\,dz$$

then, the total compaction resistance L_{cd} can be determined considering the vertical velocity effect, by the following expression:

$$L_{cd} = B \int_0^s k_1 \xi z^{n_1} dz \qquad (2.56)$$

$$\xi = \frac{1 + \lambda V_Z^K}{1 + \lambda V_0^K}$$

$$V_z = R\omega \sin\left\{\cos^{-1}\left(1 - \frac{s-z}{R}\right)\right\}$$

where V_z is the vertical speed of the plate at depth z, and a modification coefficient ξ is used to take into account the velocity effect of the terrain-wheel system constants k_1.

In the expression, L_{cd} is the value of the total land locomotion resistance of the rigid wheel. Consequently, the difference between L_{cd} and R_{cd}, which is the compaction resistance for the static amount of sinkage $s_0 - u_0$ given in Eq. (2.47), can be considered as the compaction energy due to the amount of slip sinkage s_s.

2.2.6   Effective driving force

As shown in Figure 2.8, the effective driving force T_d and the axle load W act on the central axis of a rigid wheel whilst the horizontal ground reaction B_d and the vertical reaction force N act on a point deviated from the bottom-dead-center by an eccentricity amount e_d and by a vertical distance l_d from the wheel axle. From force balance and horizontal and vertical equilibrium considerations, the effective driving force T_d and the axle load W must equal the horizontal reaction force B_d and the vertical component N respectively. T_d can now be calculated as the difference between the horizontal component of the driving force $(Q_d/R)_h$ and the compaction resistance L_{cd}, as given in the following equation:

$$T_d = B_d = \left(\frac{Q_d}{R}\right)_h - L_{cd} \qquad (2.57)$$

The amount of eccentricity e_{d0} for the no slip sinkage state can be worked out from considerations of the moment equilibrium of the vertical stress applied to the peripheral contact surface taken around the axle of the rigid wheel. Taking moments, the product of N and e_{d0} equals the integral of the product of $R \sin \theta$ and the vertical component of applied stress $pRB \cos \theta \cos(\theta - d)d\theta$ acting on an element of contact area $RB\, d\theta \cos \theta$ at an arbitrary point X on the peripheral surface of the rigid wheel. The result of doing this is:

$$Ne_{d0} = BR^2 \int_{-\theta_r}^{\theta_f} p \cos \theta \cos(\theta - \delta) \sin \theta\, d\theta$$

Substituting $N = W$ into the above equation, we get:

$$e_{d0} = \frac{BR^2}{W} \int_{-\theta_r}^{\theta_f} p \cos(\theta - \delta) \sin \theta \cos \theta\, d\theta \qquad (2.58)$$

Thence, the real eccentricity e_d of the vertical reaction force N can be modified as follows, taking into consideration the amount of slip sinkage:

where

$$\theta'_f = \cos^{-1}\left[1 - \frac{s}{R}\right] \quad \text{and} \quad \theta'_r = \theta_r$$

and

$$l_d = \frac{W e_d}{L_{cd}}$$

Using these expressions the position of the ground reaction force e_d and the value of l_d can be determined.

2.2.7 Energy equilibrium

Following the principle of conservation of energy, it is evident that the effective input energy E_1 supplied by the driving torque Q_d to the rigid wheel is equal to the sum of the individual output energy components. These components are the sinkage deformation energy E_2 required to make a rut under the rigid wheel, the slippage energy E_3 which develops at the peripheral contact part of the wheel and the effective drawbar pull energy E_4 which is required to develop an effective driving force. Energy balance requires that:

$$E_1 = E_2 + E_3 + E_4 \tag{2.59}$$

In the general case, the energy developed during the rotation of the radius vector from the entry angle θ_f to the central angle $\theta_f + \theta_r$ is calculable. As the rigid wheel rotates $R(\theta_f + \theta_r)$ and experiences a moving distance of $R(\theta_f + \theta_r)(1 - i_d)$ during this rotation, the various individual energy components can be computed as follows:

$$E_1 = Q_d(\theta_f + \theta_r) \tag{2.60}$$

$$E_2 = L_{cd} R(\theta_f + \theta_r)(1 - i_d) \tag{2.61}$$

$$E_3 = Q_d(\theta_f + \theta_r)i_d \tag{2.62}$$

$$E_4 = T_d R(\theta_f + \theta_r)(1 - i_d) \tag{2.63}$$

Next, one can note that the, value of the amount of energy developed per second can be expressed in terms of the peripheral speed $R\omega$ and the moving speed V of the rigid wheel, as follows:

$$E_1 = Q_d\omega = \left(\frac{Q_d}{R}\right)_h \frac{V}{1 - i_d} \tag{2.64}$$

$$E_2 = L_{cd} R\omega(1 - i_d) = L_{cd} V \tag{2.65}$$

$$E_3 = Q_d\omega i_d = \left(\frac{Q_d}{R}\right)_h \frac{i_d V}{1 - i_d} \tag{2.66}$$

$$E_4 = T_d R\omega(1 - i_d) = T_d V \tag{2.67}$$

The optimum effective driving force T_{dopt} during driving action is defined as the effective driving force at the optimum slip ratio i_{opt}. This occurs when the effective drawbar pull energy E_4 takes a maximum value under a constant peripheral speed $R\omega$. The tractive power efficiency E_d is defined by the following equation:

$$E_d = \frac{T_d(1 - i_d)}{[Q_d/R]_h}$$ (2.68)

2.3 AT BRAKING STATE

2.3.1 Amount of slippage

Figure 2.12 shows the distribution of the slip velocity and the amount of slippage around the peripheral surface of a rigid wheel during braking action. The skid i_b during braking action is defined for $R\omega < V$ as follows:

$$i_b = \frac{R\omega}{V} - 1$$ (2.69)

The tangential slip velocity V_s at an arbitrary point X corresponding to a central angle θ on the peripheral contact part of a rigid wheel is determined as follows. For a radius R, an angular velocity $\omega = (-d\theta/dt)$, and a moving speed of the rigid wheel V in the direction of the terrain surface the following relationship can be established.

$$V_s = R\omega - V\cos\theta$$
$$V_s = R\omega\left(1 - \frac{1}{1 + i_b}\cos\theta\right)$$ (2.70)

where θ is the angle between $\overline{OM}$ and radius vector $\overline{OX}$ and is positive for the counter-clockwise direction. V_s becomes zero at the point X_0 on the peripheral contact surface and the central angle,

$$\theta = \cos^{-1}\left(\frac{R\omega}{V}\right) = \cos^{-1}(1 - i_b)$$

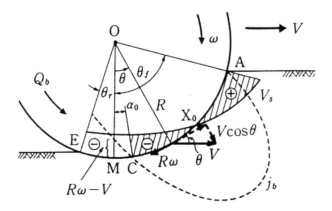

Figure 2.12. Distribution of slip velocity V_s and amount of slippage j_b during braking action ($R\omega < V$).

and the slip velocity take on positive values at the contact section $\overline{X_0 A}$ and take on negative values at the contact section $\overline{X_0 E}$.

When $i_b = \cos \theta_f - 1$ i.e. X_0 becomes coincident with the point A of the entry angle θ_f, V_s becomes negative for the whole range of contact part $\overline{AE}$.

The amount of slippage $j_b(\theta)$ is shown as the integration of V_s from the initial time $t = 0$ and a central angle $\theta = \theta_f$ at the beginning of contact with the soil to an arbitrary time $t = t$ and a central angle $\theta = \theta$. Thence, the following equation is obtained.

$$j_b = R \int_0^t \omega \left(1 - \frac{1}{1 + i_b} \cos \theta \right) dt = R \int_\theta^{\theta_f} \left(1 - \frac{1}{1 + i_b} \cos \theta \right) d\theta$$

$$= R \left\{ (\theta_f - \theta) - \frac{1}{1 + i_b} (\sin \theta_f - \sin \theta) \right\} \tag{2.71}$$

For the range of $\cos \omega_f - 1 < i_b < 0$, $j_b(\theta)$ takes the maximum value $j_b(\theta_m)$ at the point X_0 and at the central angle $\theta_m = \cos^{-1}(R\omega/V)$, where V_s becomes zero. And $j_b(\theta)$ becomes zero at the point C and at the central angle $\theta = \alpha_0$ which is derived from the following equation:

$$(\theta_f - \theta)(1 + i_b) = \sin \theta_f - \sin \theta \tag{2.72}$$

That is, the integrated area of V_s from the point A to X_0 is equal to the area of V_s integrated from the point X_0 to C.

The shear resistance $\tau(\theta)$ takes a positive value for the loading state while the amount of slippage $j_b(\theta)$ increases to a maximum value $j_b(\theta_m)$ and for the unloading state, while $j_b(\theta)$ decreases to some value $j_b(\theta_r)$ corresponding to zero shear resistance. Similarly $\tau(\theta)$ takes a negative value for the reverse traction state when $j_b(\theta)$ becomes less than $j_b(\theta_r)$. On the other hand, for the range of $i_b \leq \cos \theta_f - 1$, $j_b(\theta)$ takes a negative value and $\tau(\theta)$ also takes a negative value for the whole range of the central angle of $-\theta_r < \theta < \theta_f$.

2.3.2 Soil deformation

Figure 2.13 shows the general flow pattern beneath a towed rigid wheel in sandy soil and the position of an instantaneous center. The slip lines are largely divided into two parts for left and right hand sides from the point N on the peripheral contact area.

The forward and backward flow zones NCA and NDE are bounded by a logarithmic spiral for the transient shear zone from active to passive state and then a straight line which meets the ground surface at the angle $\pi/4 - \phi/2$ rad at the points C and D, respectively, for the passive failure zone. The parameter ϕ is the angle of internal friction of the soil.

As the instantaneous center of the towed wheel I is situated below the bottom-dead-center M, section $\widehat{NA}$, of the rim has a generally forward and downward movement and moves the soil upwards in the region NCA and therefore the soil particles slide along the rim in such a way as to produce shear resistance in the direction opposite to that of wheel revolution. Similarly section $\widehat{NE}$ of the rim has a relatively fast forward movement while the soil particles along the rim move forward slowly and therefore the shear resistance acts in the direction of wheel revolution. In general, for a towed rigid wheel, the shear resistance changes its direction at the point N on the peripheral contact surface, which is called the

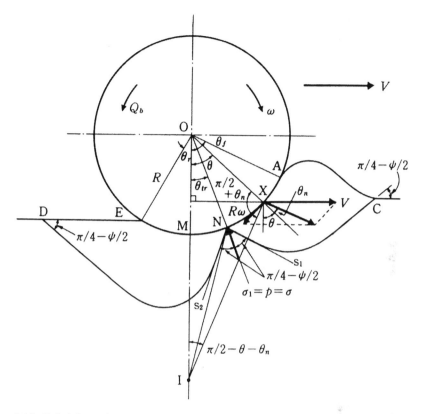

Figure 2.13. Soil deformation under a towed rigid wheel, instantaneous center I, and directions of slip line S_1, S_2 and principal stress σ_1, σ_3.

transition point. Wong et al. [19] verified experimentally for the behaviour of sandy soil beneath a towed rigid wheel – that the transition point N corresponds to the position where the two flow zones meet each other, as shown in the diagram, and the location of the transition point corresponds to the point C in the previous Figure 2.12. Point C is where the shear resistance τ becomes zero and the normal stress σ becomes the major principal stress at the interface of the soil mass. Two conjugate slip lines [20] s_1, s_2 occur forward and backward beneath the point N and intersect each other at an angle $\pm(\pi/4 - \phi/2)$ rad to the direction of the maximum principal stress σ_1 and at an angle $\pm(\pi/4 + \phi/2)$ rad to the direction of the minimum principal stress σ_3.

The resultant velocity vector of all the soil particles on the peripheral contact part of the rigid wheel always rotates around the instantaneous center I. The position of I lies below the bottom-dead-center M at a distance $\overline{OI}$ from the axle of the wheel. Using the notation shown on the diagram, the distance $\overline{OI}$ can be calculated as:

$$\overline{OI} = R\cos\theta + R\sin\theta\tan(\theta + \theta_n) = R\cos\theta + \frac{1}{\omega}(V - R\omega\cos\theta)$$

$$= \frac{V}{\omega} = \frac{R}{1 + i_b} \tag{2.73}$$

The next equation can be derived for the triangle OXI in this diagram as,

$$\frac{\overline{OI}}{\sin(\pi/2 + \theta_n)} = \frac{R}{\sin\{\pi/2 - (\theta + \theta_n)\}}$$

therefore

$$\frac{\overline{OI}}{R} = \frac{\cos\theta_n}{\cos(\theta + \theta_n)}$$

and then, a substitution of the above equation into Eq. (2.73) gives:

$$1 + i_b = \cos\theta - \sin\theta\tan\theta_n$$

$$\tan\theta_n = \frac{\cos\theta - (1 + i_b)}{\sin\theta} \tag{2.74}$$

Wong [21] concluded that the central angle $\theta = \theta_n$ corresponding to the point N could be determined by substituting $\theta_n = \pi/4 - \psi/2$ into the above equation as follows:

$$\tan\left(\frac{\pi}{4} - \frac{\psi}{2}\right) = \frac{\cos\theta_{tr} - (1 + i_b)}{\sin\theta_{tr}} \tag{2.75}$$

From the above Eq. (2.73), the instantaneous center I lies on the bottom-dead-center M at $i_b = 0\%$ and moves its location deeper below the point M with increments of skid $|i_b|$. So, the position of instantaneous center I depends on the skid i_b, or the moving speed V and the angular velocity ω of the wheel. As can be simply shown, the angle θ_n between the resultant velocity vector and the radial direction at point X becomes zero at $\angle OXI = \pi/2$ and at $\theta = \cos^{-1}(1 + i_b)$.

As will be mentioned later, the ground reaction acting on the contact part of a rigid wheel $\widehat{AN}$ develops the land locomotion resistance while another ground reaction acting on the contact part $\widehat{NE}$ develops the drag for the rigid wheel. These ground reactions to the rigid wheel apply reversely to the moving direction, and they will make the rigid wheel brake.

When a rigid wheel is locked at a skid $i_b = -100\%$, the point N approaches the point E so that the drag disappears and the slip line develops only on the curve $\widehat{NC}$. The land locomotion resistance acts as a form of bulldozing resistance to the rigid wheel.

Next, the trajectories i.e. the rolling locus of a point on the peripheral contact part of the rigid wheel during braking action will be considered. Figure 2.14 shows the rolling locus

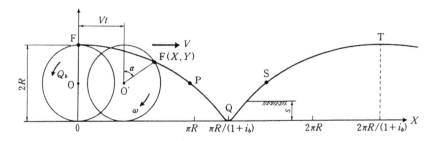

Figure 2.14. Trochoid curve during braking action ($V > R\omega$).

of an arbitrary point F on the peripheral surface of a rigid wheel for a skid i_b and $R\omega < V$ during braking action. Given that the relation between the central angle α of the point F and the time t can be expressed as $\alpha = \omega t$, the position of the axle of the rigid wheel O moves to O′ in a distance $Vt = V\alpha/\omega$.

Thence, the coordinates (X, Y) can be calculated as follows:

$$X = \frac{V\alpha}{\omega} + R\sin\alpha = R\left(\frac{\alpha}{1+i_b} + \sin a\right) \tag{2.76}$$

$$Y = R(1 + \cos\alpha) \tag{2.77}$$

The above equations imply a trochoid curve without singular points. The coordinates of point P at $\alpha = \pi/2$ are calculated as $X = R[\pi/\{2(1+i_b)\} + 1]$ and $Y = R$, the coordinates of Q at $\alpha = \pi$ are calculated as $X = R\{\pi/(1+i_b)\}$ and $Y = 0$, the coordinates of S at $\alpha = 3\pi/2$ are calculated as $X = R[3\pi/\{2(1+i_b)\} - 1]$ and $Y = R$, and the coordinates of T at $\alpha = 2\pi$ are calculated as $X = 2\pi R/(1+i_b)$ and $Y = 2R$. The gradient of the tangent to the rolling locus dY/dX is equal to the direction of the resultant velocity vector of an arbitrary point on the peripheral contact surface of the rigid wheel. This gradient can be expressed as follows:

$$\frac{dY}{dX} = -\frac{\sin\alpha}{1/(1+i_b) + \cos\alpha} \tag{2.78}$$

Then, the gradients of the tangents to the rolling locus at point P, Q, S and T can be determined as $-(1+i_b)$, 0, $+(1+i_b)$, and 0, respectively. During one revolution of a rigid wheel, there is a loss $-2\pi R i_b/(1+i_b)$ in the wheel advance and the moving distance is expressed as $2\pi R/(1+i_b)$ in conjunction with an amount of sinkage s.

Wong [15] observed the trajectories of the soil particles in a soil bin, filled with clay, when a rigid wheel was rotating during braking action.

As to his results, it was observed that the final horizontal amount of movement of a soil particle to the moving direction of the wheel decreased with depth. A soil particle located at a shallow depth in the soil bin moved first along a $\pi/4$ rad line upward to the horizontal line when the wheel approached the point. Then, the soil particle moved downward along a circular trajectory. After the pass of the wheel, the soil particle moved upward and reached its initial position due to a rebound action in the non-compressible saturated clay.

2.3.3   Force balances

The traffic pattern of a rigid wheel which is running during braking action at a constant moving speed V and at a constant angular velocity ω can be divided into two quite distinctive regimes depending on the amount of braking torque Q_b (≤ 0). One regime is the pure rolling state at $Q_b = 0$ and the other is the braking state at $Q_b < 0$.

Figure 2.15(a) shows a rigid wheel which is running in a pure rolling mode. In this case, i.e. when the braking torque becomes zero, an effective braking force $T_b = T_r$ occurs opposite to the moving direction of the wheel. T_r is called the 'rolling resistance'. It balances with the horizontal ground reaction $B_b = B_r$ which acts reversely to the moving direction of the wheel. Since the axle load W balances with the vertical ground reaction N, the direction of the resultant ground reaction (which is composed of N and $B_b = B_r$) goes through the wheel axle. The position of the application of the net ground reaction on the peripheral surface acts at a point on the wheel with an horizontal amount of eccentricity $e_b = e_r$ and a vertical distance from the axle $l_b = l_r$, as shown in the diagram. By equating forces in the horizontal and vertical directions, and by equating moments around the wheel axle the following equations may be derived:

$$W = N$$

$$T_r = B_r = \mu_r W$$

$$-Ne_r + B_r l_r = 0 \qquad (2.79)$$

Then, the horizontal amount eccentricity e_r can be calculated for a coefficient of rolling resistance μ_r as follows:

$$e_r = \mu_r l_r \qquad (2.80)$$

In a similar vein, Figure 2.15(b) shows a rigid wheel that is running in a braking state. In this case, the effective braking force T_b which is directed in the moving direction of the wheel increases when a braking torque Q_b is applied. T_b can also do external work as an actual braking force to another wheel. The horizontal ground reaction B_b acts oppositely to the moving direction of the wheel. The direction of the resultant ground reaction composed of N and B_b deviates to the left hand side of the wheel axle.

In this case, the horizontal and vertical force balances and the moment balance about the axles yield the following results:

$$W = N$$

$$T_b = B_b = \mu_b W$$

$$Q_b = -Ne_b + B_b l_b \qquad (2.81)$$

The value of the horizontal amount of eccentricity e_b can be calculated for a coefficient of braking resistance μ_b as:

$$e_b = -\frac{Q_b}{W} + \mu_b l_b \qquad (2.82)$$

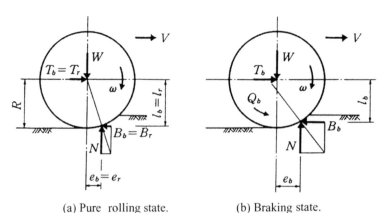

(a) Pure rolling state.   (b) Braking state.

Figure 2.15. Ground reaction acting on rigid wheel during braking action.

In this case, the actual effective braking force T_b equals the horizontal ground reaction B_b, which is given as the difference between the horizontal component of the braking force $(Q_b/R)_h$ acting reversely to the moving direction of the wheel and the compaction resistance L_{cb}, as follows:

$$T_b = \left(\frac{Q_b}{R}\right)_h - L_{cb} \tag{2.83}$$

The horizontal ground reaction B_b will be determined by both these values depending on the skid i_b.

2.3.4   Braking force

Figure 2.16 shows the contact pressure distributions that apply on the peripheral contact surface of a rigid wheel during braking action. The normal stresses $\sigma(\theta)$ have positive values for the entire contact portion of the rigid wheel. On the other hand, the shear resistances $\tau(\theta)$ applied on the contact section $\widehat{AX_0}$ have positive values in correspondence with positive amounts of slippage $j_b(\theta)$, but $\tau(\theta)$ applied on the contact section $\widehat{X_0E_0}$ turn to negative values in correspondence with negative amounts of slippage $j_b(\theta)$. The angles $\delta(\theta)$ between the resultant applied stress $p(\theta)$ and the radial direction of the wheel surface are expressed as:

$$\delta(\theta) = \tan^{-1}\left\{\frac{\tau(\theta)}{\sigma(\theta)}\right\}$$

These are positive in section $\widehat{AX_0}$, but negative in section $\widehat{X_0E}$.

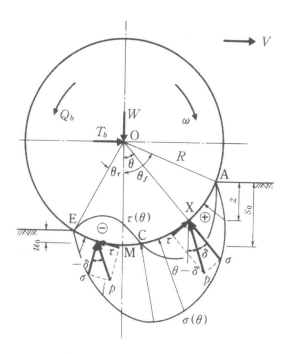

Figure 2.16. Contact pressure distributions applied on peripheral surface of rigid wheel during braking action.

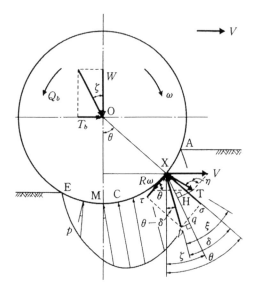

Figure 2.17. Component of a rolling locus $\Delta d(\theta)$ in the direction of applied stress during braking action.

At the point X_0 at which the shear resistance $\tau(\theta)$ and the slip velocity V_s become zero, and at which the amount of slippage j_b takes a maximum value, the applied normal stress $\sigma(\theta)$ becomes the maximum principal stress σ_1 and the direction of the resultant stress $p = \sigma(\theta)$ is coincident with the radial direction of the rigid wheel.

The amount of sinkage z at an arbitrary point X on the peripheral surface, the amount of sinkage s_0 at the bottom-dead-center M and the amount of rebound u_0 at the point E can all be expressed as shown in Eqs. (2.32) ~ (2.34).

The length of rolling locus $l = l(\alpha)$ can be derived from Eqs. (2.76) and (2.77). Substituting the relation $\theta = \pi - \alpha$ into $l(\alpha)$, the length of the trajectory $l = l(\theta)$ is given by integrating the elemental locus from $\theta = \theta$ to $\theta = \theta_f$ as follows:

$$l(\theta) = R \int_{\theta}^{\theta_f} \left\{ \left(\frac{1}{1+i_b} \right)^2 - \frac{2\cos\theta}{1+i_b} + 1 \right\}^{\frac{1}{2}} d\theta \tag{2.84}$$

As shown in Figure 2.17, the direction of the resultant force between the effective braking force T_b and the axle load W is given by the angle $\zeta = \tan^{-1}(T_b/W)$ to the vertical axis. In this case, the relation between contact pressure $q(\theta)$ and soil deformation $d(\theta)$ in the direction of the resultant force agrees well with the plate loading, unloading and sinkage relationships. Here, $q(\theta)$ is the component of the resultant applied stress $p(\theta)$ to the direction of the angle ζ to the vertical axis. $d(\theta)$ is the component of the rolling locus $l(\theta)$ to the same direction of the resultant force. As $\overline{XT}$ is the element of the rolling locus in the same direction of the resultant velocity vector, $d(\theta)$ may be calculated as the integral of $\overline{XH}$ from $\theta = \theta$ to $\theta = \theta_f$, which is the component of $\overline{XT}$ in the direction of the angle ζ to the vertical axis, as follows:

$$d(\theta) = R \int_{\theta}^{\theta_f} \left\{ \left(\frac{1}{1+i_b}\right)^2 - \frac{2\cos\theta}{1+i_b} + 1 \right\}^{\frac{1}{2}} \times \sin\left(\theta - \varsigma + \tan^{-1}\frac{\sin\theta}{1+i_b - \cos\theta}\right) d\theta$$

(2.85)

$$\eta = \tan^{-1}\frac{V\sin\theta}{Rw - V\cos\theta} = \tan^{-1}\frac{\sin\theta}{1+i_b - \cos\theta}$$

$$\xi = \delta + \eta - \pi/2, \quad \angle HXT = \eta - (\pi/2 - \theta + \zeta)$$

$$\Delta d(\theta) = \overline{XH} = \overline{XT}\cos(\angle HXT) = \Delta l(\theta)\sin(\theta - \varsigma + \eta)$$

$$q = p\cos(\zeta - \theta + \zeta)$$

The velocity vector V_p in the direction of the stress $q(\theta)$ applied to the peripheral surface of a rigid wheel can be calculated as the component of the resultant velocity between $R\omega$ and V, as follows:

$$V_p = \left\{ (R\omega - V\cos\theta)^2 + (V\sin\theta)^2 \right\}^{\frac{1}{2}} \times \sin\left(\theta - \varsigma + \tan^{-1}\frac{\sin\theta}{1+i_b - \cos\theta}\right) \quad (2.86)$$

In this case, the plate loading and unloading test should be carried out allowing for the 'size effect' of the contact length of the wheel $b = R(\sin\theta_f + \sin\theta_r)$ and the 'velocity effect' of the loading speed V_p. Thereafter, the relationship between the resultant applied stress $p(\theta)$, the stress component $q(\theta)$ and the soil deformation $d(\theta)$ can be determined as:

For $\theta_{max} \leq \theta \leq \theta_f$

$$p(\theta) = \frac{k_1\xi\{d(\theta)\}^{n_1}}{\cos(\varsigma - \theta + \delta)}$$

For $-\theta_r < \theta < \theta_{max}$

$$p(\theta) = \frac{k_1\xi\{d(\theta_{max})\}^{n_1} - k_2\{d(\theta_{max}) - d(\theta)\}^{n_2}}{\cos(\varsigma - \theta + \delta)}$$

(2.87)

$$\xi = \frac{1 + \lambda V_p^k}{1 + \lambda V_0^k}$$

$$k_1 = \frac{k_{c1}}{b} + k_{\phi1} \quad \text{and} \quad k_2 = \frac{k_{c2}}{b} + k_{\phi2}$$

where θ_{max} is the central angle corresponding to the maximum stress $q_{max} = q(\theta_{max})$. The coefficients k_{c1}, $k_{\phi1}$ and k_{c2}, $k_{\phi2}$ and the indices n_1 and n_2 in this case need to determined from the quasi-static plate loading and unloading test which is carried out at a loading speed V_0. The other coefficients λ and the index κ need to be determined by use of a dynamic plate loading test carried out at a loading speed V_p.

Thence, the distribution of normal stress $\sigma(\theta)$ can be calculated as:

$$\sigma(\theta) = p(\theta)\cos\delta$$

(2.88)

Similarly, the distribution of shear resistance $\tau(\theta)$ may be calculated by substituting the amount of slippage $j_b(\theta)$ in Eq. (2.71) into the previously presented Janosi-Hanamoto equation, as follows:

For a first mode case where $\cos\theta_f - 1 < i_b < 0$:

For a traction state of $0 < j_b(\theta) < j_p$,

$$\tau(\theta) = \{c_a + \sigma(\theta)\tan\phi\}[1 - \exp\{-aj_b(\theta)\}]$$

For a untraction state of $j_q < j_b(\theta) < j_p$,

$$\tau(\theta) = \tau_p - k_0\{j_p - j_b(\theta)\}^{n_0}$$

For a reciprocal traction state of $j_b(\theta) \leq j_q$,

$$\tau(\theta) = -\{c_a + \sigma(\theta)\tan\phi\}(1 - \exp[-a\{j_q - j_b(\theta)\}]) \qquad (2.89)$$

where j_p is the maximum amount of slippage, τ_p is the shear resistance at j_p, and j_q is the amount of slippage when τ takes on a value of zero during the untraction state.

For a first mode case where $i_b < \cos\theta_f - 1$:

$$\tau(\theta) = -\{c_a + \sigma(\theta)\tan\phi\}[1 - \exp\{aj_b(\theta)\}] \qquad (2.90)$$

The above equations can be applied for loose sandy soil or weak clayey soft ground, but another equation [12] is better used in the case of hard compacted sandy ground.

Taking a next step, the axle load W, the braking torque Q_b, and the apparent effective braking force T_{b0} can be related to the contact pressure distribution $\sigma(\theta)$ and $\tau(\theta)$, as follows:

$$W = BR\int_{-\theta_r}^{\theta_f} \{\sigma(\theta)\cos\theta + \tau(\theta)\sin\theta\}\,d\theta \qquad (2.91)$$

$$Q_b = BR^2\int_{-\theta_r}^{\theta_f} \tau(\theta)\,d\theta \qquad (2.92)$$

$$T_{bo} = BR\int_{-\theta_r}^{\theta_f} \{\tau(\theta)\cos\theta - \sigma(\theta)\sin\theta\}\,d\theta \qquad (2.93)$$

Thence, the braking force can be computed as the value of Q_b/R.

In the above equations, $Q_b \leq 0$ and $T_{b0} < 0$, and the apparent effective braking force T_{b0} can be expressed as the difference between the drag $T_{hb}(<0)$ and the compaction resistance $R_{cb}(>0)$, as follows:

$$T_{bo} = T_{hb} - R_{cb} \qquad (2.94)$$

where

$$T_{hb} = BR\int_{-\theta_r}^{\theta_f} \tau(\theta)\cos\theta\,d\theta \qquad (2.95)$$

$$R_{cb} = BR\int_{-\theta_r}^{\theta_f} \sigma(\theta)\sin\theta\,d\theta \qquad (2.96)$$

Further, the horizontal component of the braking force $(Q_b/R)_h$ may be realised as the sum of the actual effective braking force T_b and the total compaction resistance L_{cb} which is the land locomotion resistance calculated for the amount of slip sinkage.

2.3.5   Compaction resistance

Figure 2.18 shows the rolling motion of a rigid wheel during braking action on a ground surface.

To calculate the amount of slip sinkage s_s of a wheel at a point X on the ground surface, it is necessary to determine the amount of slippage j_s at a point X. The moving distance $\overline{OO'}$ of the wheel during the time t is given from Eq. (2.49) as follows:

$$\overline{OO'} = Vt = \frac{R\omega t}{1 + i_b} = \frac{R(\theta_f - \theta)}{1 + i_b} \tag{2.97}$$

The transit time t_e during which a rigid wheel passes over a point X is given as the time when the distance $\overline{OO'}$ reaches the contact length of wheel $\overline{OO''}$ and the central angle of the radius vector $\overline{O''F}$ becomes $\theta = \theta_e$ as follows:

$$\overline{OO''} = R(\sin\theta_f + \sin\theta_r)$$
$$\theta_e = \theta_f - (1 + i_b)(\sin\theta_f + \sin\theta_r) \tag{2.98}$$

therefore

$$t_e = \frac{1}{\omega}(1 + i_b)(\sin\theta_f + \sin\theta_r) \tag{2.99}$$

is obtained.

Since the amount of slippage of a rigid wheel during one revolution of the wheel is equivalent to $-2\pi R i_b/(1 + i_b)$, the amount of slippage of a soft ground at a point X can be

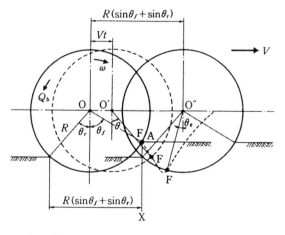

Figure 2.18. Rolling motion of a towed rigid wheel for point X on ground surface.

calculated as the ratio of t_e to $2\pi/\omega$ as follows:

$$j_s = \frac{-2\pi R i_b}{1 + i_b} \cdot \frac{R(\sin\theta_f + \sin\theta_r)}{2\pi R/(1 + i_b)} = -R i_b(\sin\theta_f + \sin\theta_r) \tag{2.100}$$

Considering a small interval of the central angle $(\theta_f + \theta_r)/N$ and a small interval of the amount of slippage j_s/N for a very small time interval t_e/N, the vertical component of the resultant applied stress p can be calculated as $p(\theta_n)\cos(\theta_n - \delta_n)$ for the nth interval of the central angle θ_n. Then, substituting this into Eq. (2.48), the amount of slip sinkage s_s can be developed as the sum of all the elemental amounts of slip sinkage as follows:

$$s_s = c_0 \sum_{n=1}^{N} \{p(\theta_n)\cos(\theta_n - \delta_n)\}^{c_1} \left\{ \left(\frac{n}{N}j_s\right)^{c_2} - \left(\frac{n-1}{N}j_s\right)^{c_2} \right\} \tag{2.101}$$

where

$$\theta_n = \frac{n}{N}(\theta_f + \theta_r)$$

Further, the total sinkage i.e. the rut depth of the rigid wheel s can be calculated given the amount of static sinkage s_0, the amount of slip sinkage s_s and the amount of rebound u_0, as follows:

$$s = s_0 - u_0 + s_s \tag{2.102}$$

The product of the compaction resistance L_{cb} applied in front of the rigid wheel and the moving distance $2\pi R/(1 + i_b)$ during one revolution of the wheel can be equated to the work in making the rut. This work component can be calculated as the integration of the contact pressure p acting on a plate of length $2\pi R/(1 + i_b)$ and of B from a depth $z = 0$ to the total amount of sinkage $z = s$, as follows:

$$\frac{2\pi R L_{cd}}{1 + i_b} = \frac{2\pi R B}{1 + i_b} \int_0^s p\,dz$$

Thence the compaction resistance L_{cb} can be determined as follows, considering the vertical velocity effect:

$$L_{cb} = k_1 \xi B \int_0^s z^{n_1}\,dz \tag{2.103}$$

where

$$\xi = \frac{1 + \lambda V_z^k}{1 + \lambda V_0^k}$$

$$V_z = R\omega \sin\left\{\cos^{-1}\left(1 - \frac{s-z}{R}\right)\right\}$$

Additionally, L_{cb} is the value of the total land locomotion resistance to the rigid wheel's passage. Hence, the difference between L_{cb} and R_{cb} – which is the compaction resistance for the static amount of sinkage $s_0 - u_0$ given in Eq. (2.96) – can be considered as the compaction resistance due to the amount of slip sinkage s_s.

2.3.6   Effective braking force

As shown in Figure 2.15, the effective braking force T_b and the axle load W act on the central axis of a rigid wheel. Likewise, the horizontal ground reaction B_b and the vertical reaction N act on a point deviated from the bottom-dead-center by an eccentricity amount e_b and by a vertical distance l_b from the wheel axle. From force balance considerations, T_b and W equal B_b and N respectively. T_b can be calculated as the difference between the horizontal component of the braking force $(Q_b/R)_h$ and the compaction resistance L_{cb}, as in the following equation:

$$T_b = B_b = \left(\frac{Q_b}{R}\right)_h - L_{cb} \tag{2.104}$$

The amount of eccentricity e_{b0} for the no slip sinkage state is given as follows, considering the moment equilibrium of vertical stress applied to the peripheral contact surface around the axle of the rigid wheel. That is, the product of N and e_{b0} equals the integral of the product of $R \sin \theta$ and the vertical component of the applied stress times the elemental contact area as predicted for the driving state.

Then, the amount of eccentricity e_{b0} can be calculated as follows:

$$e_{b0} = \frac{BR^2}{W} \int_{-\theta_r}^{\theta_f} p \cos(\theta - \delta) \sin \theta \cos \theta \, d\theta$$

Following this, the real eccentricity e_b of the vertical ground reaction N can be modified as:

$$e_b = \frac{(\sin \theta_f' + \sin \theta_r')(R \sin \theta_r + e_{b0})}{\sin \theta_f + \sin \theta_r} - R \sin \theta_r \tag{2.105}$$

$$\theta_f' = \cos^{-1}\left[1 - \frac{s}{R}\right] \quad \text{and} \quad \theta_r' = \theta_r$$

and

$$l_b = \frac{W e_b}{L_{cb}}$$

Finally, the position of ground reaction force, e_b and l_b can be determined.

2.3.7   Energy equilibrium

Applying the conservation of energy principle, it is clear that the effective input energy E_1 supplied by the braking torque Q_b to a rigid wheel is equal to the sum of the individual output energy components. These components are the sinkage deformation energy E_2 required to make a rut under the rigid wheel, the slippage energy E_3 which develops on the peripheral contact part of the wheel and the effective braking force energy E_4.

In this case, the energy developed during the rotation of the radius vector from the entry angle θ_f to the central angle $\theta_f + \theta_r$ can be considered. As the rigid wheel rotates $R(\theta_f + \theta_r)$ and the moving distance becomes $R(\theta_f + \theta_r)/(1 + i_b)$, each component energy factor can be quantitatively defined as follows:

$$E_1 = Q_b(\theta_f + \theta_r) \tag{2.106}$$

$$E_2 = \frac{L_{cb}R(\theta_f + \theta_r)}{1 + i_b} \tag{2.107}$$

$$E_3 = \frac{Q_b(\theta_f + \theta_r)i_b}{1 + i_b} \tag{2.108}$$

$$E_4 = \frac{T_bR(\theta_f + \theta_r)}{1 + i_b} \tag{2.109}$$

Next, one can note that the value of the amount of energy developed per second can be expressed in terms of the peripheral speed $R\omega$ and the moving speed V of the wheel, as follows:

$$E_1 = Q_b\omega = \left(\frac{Q_b}{R}\right)_h V(1 + i_b) \tag{2.110}$$

$$E_2 = \frac{L_{cb}R\omega}{1 + i_b} = L_{cb}V \tag{2.111}$$

$$E_3 = \frac{Q_b\omega i_b}{1 + i_b} = \left(\frac{Q_b}{R}\right)_h i_b V \tag{2.112}$$

$$E_4 = \frac{T_bR\omega}{1 + i_b} = T_bV \tag{2.113}$$

The optimum effective braking force T_{bopt} is defined as the effective braking force at the optimum skid i_{bopt} when the effective input energy $|E_1|$ takes a maximum value. Additionally, the braking power efficiency E_b can be defined as:

$$E_b = \frac{T_b}{(Q_b/R)_h(1 + i_b)} \tag{2.114}$$

2.4   SIMULATION ANALYSIS

Figure 2.19 presents a flow chart that can be used to calculate the tractive performance of a driven rigid wheel or the braking performance of a towed rigid wheel when it is running on a flat soft ground. First of all, the various wheel dimensions – such as the axle load W, the radius R, the width B of the rigid wheel, the peripheral velocity $R\omega$ or the wheel velocity V are required to be provided as a set of input data.

Following this, the terrain-wheel system constants – such as the coefficients k_{c1}, $k_{\phi1}$, and k_{c2}, $k_{\phi2}$ and the indices n_1, n_2, and the loading rate V_0 measured from the quasi-static plate loading and unloading test, the coefficient α and the index κ measured from the dynamic plate loading test, the soil constants c_a, $\tan\phi$ and a measured from the plate traction test, as well as the coefficient c_0, the indices c_1, c_2 measured from the slip sinkage test are also required as another set of input data.

For the rest condition, the distributions of normal stress $\sigma(\theta)$, shear resistance $\tau(\theta)$, resultant applied stress $p(\theta)$ and friction angle $\delta(\theta)$ can be calculated for a given entry angle

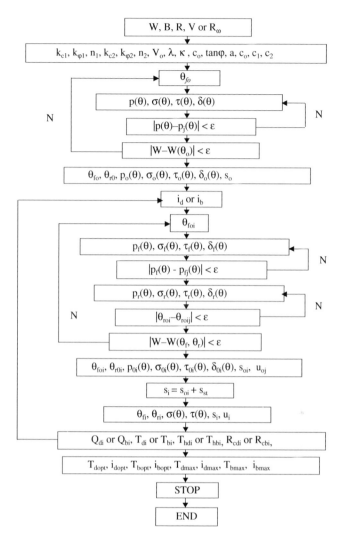

Figure 2.19. Simulation flow chart.

θ_{f0} and for a given exit angle θ_{r0}. Thence, to determine the real values of θ_{f0} and θ_{r0}, the distributions of $\sigma(\theta)$, $\tau(\theta)$, $p(\theta)$ and $\delta(\theta)$ can be iteratively calculated by means of the two division method until the vertical equilibrium Eq. (2.91) can be precisely satisfied.

At driving or braking state for a given slip ratio i_d or skid value i_b, the following two calculations need to be undertaken so that the entry angle θ_{foi} may be determined. For the forward peripheral contact part $\widehat{AM}$, the distribution of $\sigma_f(\theta)$ calculated from Eq. (2.39) or (2.88), $\tau_f(\theta)$ calculated from Eq. (2.41) or (2.90), which can be calculated from $p_f(\theta)$ given in Eq. (2.38) or (2.87) are calculated repeatedly until the real distribution of $p_f(\theta)$ is determined. After that, for the backward peripheral contact part $\widehat{ME}$, the distribution of $\sigma_r(\theta)$, $\tau_r(\theta)$, and $\delta_r(\theta)$ calculated from $p_r(\theta)$ are computed until the exit angle θ_{roi} is determined.

Next, calculations should be done for the given entry angle θ_{foi} until the vertical equilibrium Eq. (2.42) or (2.91) are satisfied precisely for the given slip ratio i_d or skid value i_b. Then, the final values of θ_{foi} and θ_{roi}, the final distribution of $p_f(\theta)$ and $\sigma_f(\theta)$, $\tau_f(\theta)$, $\delta_f(\theta)$, and $p_r(\theta)$ and $\sigma_r(\theta)$, $\tau_r(\theta)$, and $\delta_r(\theta)$, the final amount of sinkage s_{0i} and the final amount of rebound u_0 can be determined. Further to this, the total amount of sinkage s_i may be calculated from Eq. (2.55) or (2.102) else can be determined from the amount of slip sinkage s_{si} given in Eq. (2.54) or (2.101). Then, the driving or braking torque Q_{di} or Q_{bi} can be calculated from Eq. (2.43) or (2.92), the driving force Q_{di}/R or the braking force Q_{bi}/R, the compaction resistance L_{cdi} or L_{cbi} can be calculated using Eq. (2.56) or (2.103), the amount of eccentricity e_{di} or e_{bi} can be calculated from Eq. (2.58) or (2.105), the tractive or braking power efficiency E_{di} or E_{bi} can be calculated from Eq. (2.68) or (2.114) and several energy values E_{1_i}, E_{2_i}, E_{3_i} and E_{4_i} can be calculated from Eq. (2.60) $\sim$ (2.67) or (2.106) $\sim$ (2.113) and can be determined for all the slip ratios i_d or skids i_d. Additionally, the optimum slip ratio i_{dopt} or the optimum skid i_{bopt} and the optimum effective driving or braking force T_{dopt} or T_{bopt} can be determined. Finally, the relations between Q_d/R–i_d, T_d–i_d, s–i_d, θ_{df}, θ_{dr}–i_d, e_d–i_d, E_d–i_d and Q_b/R–i_b, T_b–i_b, s–i_b, θ_{bf}, θ_{br}–i_b, e_b–i_b, E_b–i_b, and, E, E_2, E_3, E_4–i_d or i_b, and the distributions of the normal stress $\sigma(\theta)$ and the shear resistance $\tau(\theta)$ for all the slip ratios i_d or skids i_d can be graphically developed and portrayed by use of an ordinary microcomputer.

2.4.1 Driving state

To give an example of these computational processes and to validate them, the tractive performance of a driven rigid wheel with an axle load $W = 1.52$ kN, and geometrical properties radius $R = 16$ cm and width $B = 9.5$ cm has been simulated for a wheel running on a weak soft soil ground with a peripheral speed $R\omega = 7.07$ cm/s. The analytical simulation results have then been contrasted and verified by comparison with detailed experimental test data. All the terrain-wheel system constants for the experimental situation are given in Table 2.1.

The sandy soil in the experiment was dried in air and had a water content $w = 2.38\%$, a specific gravity $G_s = 2.66$, an average grain size $D_{50} = 0.78$ mm and a coefficient of uniformity $U_c = 12.0$. The size of soil bin was 120 cm in length, 10 cm in width and 35 cm in depth. The sandy soil was filled uniformly into the two dimensional soil bin by means of a free fall method that employed a 35 cm drop height. The initial density of the sandy

Table 2.1. Terrain-wheel system constants.

Plate loading and unloading test		
$k_{c1} = 14.46$ N/cm$^{n_1+1}$	$k_{c2} = 48.10$ N/cm$^{n_2+1}$	
$k_{\phi_1} = 4.95$ N/cm$^{n_1+2}$	$k\phi = 36.47$ N/cm$^{n_2+2}$	
$n_1 = 0.809$	$n_2 = 0.757$	
$\lambda = 0.18$		
$\kappa = 1.35$	$V_0 = 0.035$ cm/s	
Plate traction and slip sinkage test		
$c_a = 0$ kPa	$\tan\phi = 0.423$	$a = 1.76$ 1/cm
$c_0 = 9.738 \times 10^{-6}$	$c_1 = 2.065$	$c_2 = 1.074$

soil was prepared to be $\gamma = 1.52 \, \text{mg/m}^3$. The actual traction test on the rigid wheel was executed under conditions of plane strain with the driving torque Q_d, the effective driving force T_d, and the total amount of sinkage s being measured directly.

Figure 2.20 shows the experimental relationships that exist between the driving force Q_d/R, the effective driving force T_d and the slip ratio i_d. Q_d/R increases gradually with increments in i_d and increases rapidly from $i_d \cong 55\%$. T_d also increases gradually with i_d and reaches a maximum value $T_{dmax} = 0.127 \, \text{kN}$ at $i_{dm} = 21\%$. After that, T_d takes a negative value at $i_d = 43\%$ and then decreases rapidly at slip ratios more than about 55%. As a consequence, the rigid wheel cannot move any more – without external pushing – at this slip ratio. In this case, the optimum effective driving force T_{dopt} is calculated as 0.097 kN at an optimum slip ratio $i_{dopt} = 10\%$. The analytical simulation results agree well with the experimental test data, as shown in this diagram indicated by the o and • plotted values.

Figure 2.21 shows the experimentally derived relationship between the total amount of sinkage s and the slip ratio i_d. The sinkage s increases gradually with increments of i_d due to increasing amount of slip sinkage till $i_d \cong 55\%$. After this the sinkage increases rapidly. The analytical results agree well with the experimental test result in this case as shown by the computed value • marks on the figure.

Figure 2.22 shows the relationship between the amount of eccentricity e_d of vertical reaction force and the slip ratio i_d. The parameter e_d increases parabolically with increasing i_d, after taking a minimum value $e_{dmin} = 4.2 \, \text{cm}$ at $i_d \cong 0\%$, due to the increasing amount of slip sinkage.

Figure 2.23 shows the relationships between the entry angle θ_f, the exit angle θ_r and the slip ratio i_d. The parameter θ_f increases parabolically with increments of i_d after taking a minimum value $\theta_{fmin} = 0.629 \, \text{rad}$ at $i_d \cong 0\%$. In contrast, the value of θ_r increases gradually with i_d. It rises from $\theta_r = 0.175 \, \text{rad}$ at $i_d \cong 0\%$ and decreases rapidly after reaching a maximum value $\theta_{rmax} = 0.244 \, \text{rad}$.

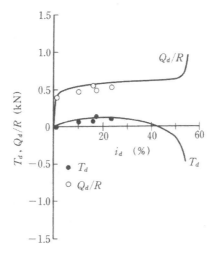

Figure 2.20. Relationships between driving force Q_d/R, effective driving force T_d and slip ratio i_d.

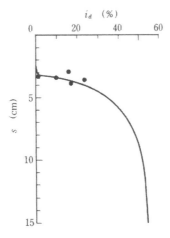

Figure 2.21. Relationship between total amount of sinkage S and slip ratio i_d during driving action.

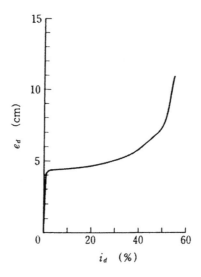

Figure 2.22. Relationship between amount of eccentricity e_d and slip ratio i_d during driving action.

Figure 2.24 shows the relationships between the various energy values E_1, E_2, E_3, E_4 and the slip ratio i_d. The effective input energy E_1 increases gradually with increments in i_d and increases rapidly from $i_d \cong 55\%$. The sinkage deformation energy E_2 increases parabolically with i_d from an initial value of 0.934 kNcm/s at $i_d \cong 0\%$. The slippage energy E_3 increases almost linearly with i_d but increases rapidly from $i_d \cong 55\%$. The effective drawbar pull energy E_4 increases with i_d and reaches a maximum value $E_{4max} = 2.359$ kNcm/s at $i_d = 10\%$. After that, E_4 decreases almost parabolically till it develops negative values.

Figure 2.25 shows the relationship between the tractive power efficiency E_d and the slip ratio i_d. E_d decreases almost hyperbolically with i_d from a maximum value $E_{dmax} = 68.1\%$ at $i_d \cong 0\%$ and takes on negative values from $i_d = 43\%$. Figure 2.26 shows the distributions of normal stress $\sigma(\theta)$ and shear resistance $\tau(\theta)$ at $i_{dopt} = 10\%$. The shape of these stress

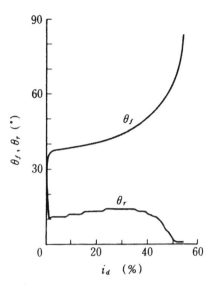

Figure 2.23. Relationships between entry angle θ_f, exit angle θ_r and slip ratio i_d during driving action.

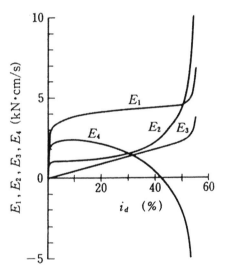

Figure 2.24. Relationship between energy values E_1, E_2, E_3, E_4, and slip ratio i_d during driving action.

distribution agrees well with the experimental test data presented in a paper published by Onafeko et al. [22]. In this case, a maximum value of normal stress $\sigma_{max} = 156.2$ kPa is obtained at $\theta_N = 0.300$ rad and $\theta_N/\theta_f = 0.450$.

Figure 2.27 shows the relationship between the angle ratio θ_N/θ_f to obtain the maximum normal stress and the slip ratio i_d. The ratio θ_N/θ_f increases slightly with increasing values of i_d, and the constants in Eq. (2.17), $a = 0.391$ and $b = 1.82 \times 10^{-3}$ are obtained. It is clear from this graph that the position showing the maximum normal stress shifts forward

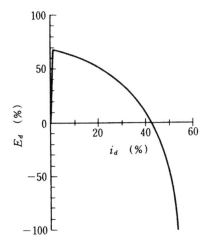

Figure 2.25. Relationship between tractive efficiency E_d and slip ratio i_d.

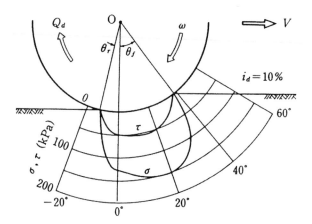

Figure 2.26. Distributions of contract pressure; normal stress σ and shear resistance τ, at the optimum slip ratio $i_{dopt} = 10\%$.

39.1% of the entry angle and that it increases slightly with increasing slip ratio. Hiroma et al. [23] also observed the same experimental results.

2.4.2 Braking state

In a similar manner to the foregoing and as a further example, the braking performance of a towed rigid wheel with axial load $W = 1.52$ kN, radius $R = 16$ cm, width $B = 9.5$ cm running on a soft sandy soil ground at a wheel speed $V = 7.07$ cm/s has been simulated. Likewise the mathematical simulation results have been verified by comparison with experimental test data. All the terrain-wheel system constants and the soil properties are the same as given in the previous session 2.4.1 example for the driving state.

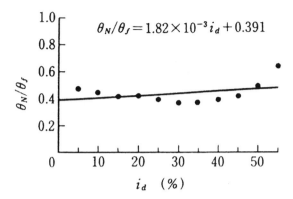

Figure 2.27. Relationship between angle ratio θ_N/θ_f and slip ratio i_d during driving action.

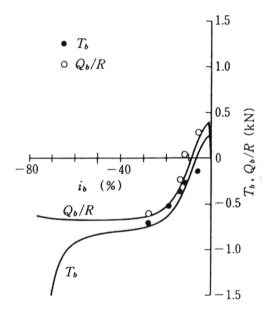

Figure 2.28. Relationship between braking force Q_b/R, effective braking force T_b and skid i_b during braking action.

The experimental towing test of the rigid wheel was executed under conditions of plane strain and the braking torque Q_b, the effective braking force T_b and the total amount of sinkage s were measured directly.

Figure 2.28 shows the relationship between the braking force Q_b/R, the effective braking force T_b and the skid parameter i_b. Q_b/R decreases rapidly with increments of $|i_b|$ and takes a zero value at $i_b \cong -9\%$. Afterwards, $|Q_b/R|$ increases parabolically with $|i_b|$. It reaches a maximum value $|Q_b/R|_{max} = 0.674$ kN at $i_b = -52\%$. $|T_b|$ also decreases rapidly with increasing values of $|i_b|$ and it reaches zero at $i_b = -7\%$. Afterwards, $|T_b|$ increases parabolically with $|i_b|$. It increases rapidly at skid values past $|i_b| \cong 70\%$ due to increasing land locomotion resistance.

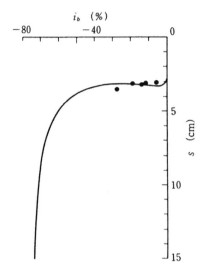

Figure 2.29. Relationship between total amount of sinkage S and skid i_b during braking action.

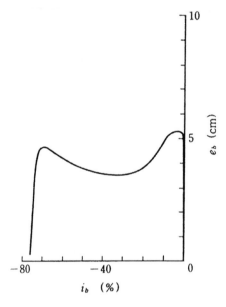

Figure 2.30. Relationships between amount of eccentricity e_b of vertical reaction and skid i_b during braking action.

In this situation, the optimum effective braking force T_{bopt} is calculated as $-0.737\,\text{kN}$ at an optimum slip ratio $i_{bopt} = -27\%$. The analytical simulation results compare well with the experimental test data, as illustrated by the o and • calculated values.

Figure 2.29 shows the relationship between the total amount of sinkage s and skid i_b. Initially, the sinkage s is almost constant with increasing $|i_b|$ due to the balance of the

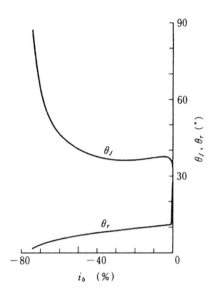

Figure 2.31. Relationships between entry angle θ_f, exit angle θ_r and skid i_b during braking action.

decreasing amount of static sinkage and the increasing amount of slip sinkage. However the sinkage begins to increase rapidly with $|i_b|$ at skid values greater than $|i_b| \cong 70\%$ due to increasing amounts of slip sinkage. The analytical results agree well with the experimental test results as shown by the • marks in the diagram.

Figure 2.30 shows the relationship between the amount of eccentricity e_b of the vertical reaction and skid i_b. The value of e_b initially decreases gradually with increments of $|i_b|$ after taking the maximum value $e_{bmax} = 5.3$ cm at $i_b = -3\%$ due to the decreasing amount of static sinkage. After this it takes a minimum value $e_{bmin} = 3.5$ cm at $i_b = -33\%$. Afterwards, e_b increases gradually with $|i_b|$ until it reaches an upper peak value $e_b = 4.7$ cm at $i_b = -69\%$.

Figure 2.31 shows the relationship between the entry angle θ_f, the exit angle θ_r and skid i_b. The value of θ_f increases parabolically with increasing $|i_b|$ after taking a minimum value $\theta_{fmin} = 0.630$ rad at $i_b \cong -1\%$. On the other hand, θ_r decreases gradually with increasing values of $|i_b|$ from a maximum value $\theta_{rmax} = 0.189$ rad at $i_b \cong -3\%$.

Figure 2.32 shows the relationships between the several energy components E_1, E_2, E_3, E_4 and the skid parameter i_b. The effective input energy E_1 decreases rapidly with increments in $|i_b|$ and reaches zero at $i_b = -9\%$. Afterwards, $|E_1|$ increases parabolically with $|i_b|$ and reaches a maximum value $|E_1|_{max} = 3.197$ kNcm/s at $|i_b| = 27\%$. After that, $|E_1|$ decreases gradually to zero. In contrast, the sinkage deformation energy E_2 increases parabolically with $|i_b|$, after reaching a minimum value $E_{2min} = 0.810$ kNcm/s at $i_b \cong -33\%$. The slippage energy E_3 increases almost linearly with increments of $|i_b|$ past $|i_b| \cong 9\%$. The effective braking force energy E_4 decreases rapidly with increases in $|i_b|$ and reaches zero at $i_b \cong -7\%$. Afterwards, $|E_4|$ increases rapidly with $|i_b|$. It increases suddenly at skid values of more than $|i_b| \cong 70\%$.

Computationally, the optimum braking force $(Q_b/R)_{opt} = -0.619$ kN, the optimum effective braking force $T_{bopt} = -0.737$ kN, the total amount of sinkage $s = 3.1$ cm, the amount of eccentricity of the vertical reaction $e_b = 3.6$ cm, the braking power efficiency $E_b = 163\%$,

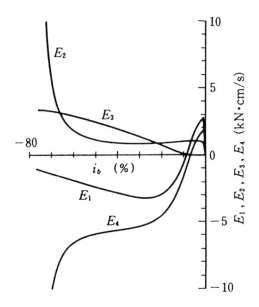

Figure 2.32. Relationships among energy values E_1, E_2, E_3, E_4 and skid i_b during braking action.

the entry angle $\theta_f = 0.635$ rad and the exit angle $\theta_r = 0.158$ rad can be obtained for the optimum skid $i_{bopt} = -27\%$ that might be used to maximize the effective input energy E_1.

Figure 2.33 shows the distributions of normal stress $\sigma(\theta)$ and shear resistance $\tau(\theta)$ at values of $i_b = -10$, -20 and -30%. The shape of these stress distribution agrees well with experimental results published by Onafeko et al. [22], Krick [24] and Ogaki [25]. The distribution of the shear resistance $\tau(\theta)$ at $i_b = -10\%$ shows a change from a positive value to a negative one at a central angle $\theta = 0.271$ rad. In contrast the distributions of $\tau(\theta)$ at $i_b = -20\%$ and -30% are always negative.

In these cases, the maximum values of normal stress $\sigma_{max} = 173.6$, 166.5 and 165.4 kPa are obtained at central angles $\theta_N = 0.277$, 0.238 and 0.096 rad and $\theta_N/\theta_f = 0.425$, 0.375 and 0.150, respectively.

Figure 2.34 shows the relationship between the angle ratio θ_N/θ_f to obtain maximum normal stress and skid i_b. The ratio θ_N/θ_f decreases gradually with increasing values of $|i_b|$. The ratio takes a minimum value $(\theta_N/\theta_f)_{min} = -0.209$ at $i_b = -40\%$ and then increases slightly with $|i_b|$. This tendency can be compared to the experimental test data presented by Oida et al. [26]. In this case, it is clearly shown that the position showing the maximum normal stress shifts forward 24.0% of the entry angle at an optimum slip value $i_{bopt} = -27\%$.

2.5   SUMMARY

In this chapter, we have studied the simplest of the machine-terrain interaction problems namely that of predicting the behaviour of a, loaded, rigid cylindrical drum operating upon a compressible, and potentially yielding, medium. The prediction is made by developing mathematical models of the system.

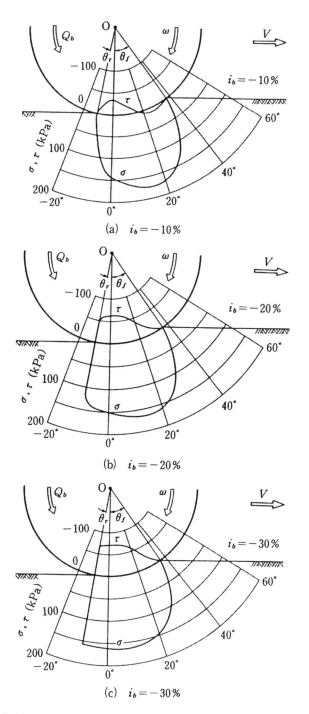

Figure 2.33. Distributions of contact pressure; normal stress σ and shear resistance τ at (a) $i_b = -10\%$, (b) $i_b = -20\%$ and (c) $i_b = -30\%$ during braking action.

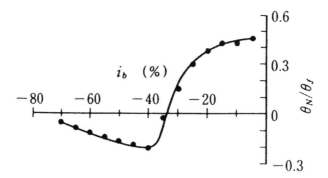

Figure 2.34. Relationship between angle ratio θ_N/θ_f and skid i_b during braking action.

Up to now a number of different simple drum/terrain systems have been modelled. These include:
– A static, loaded, cylindrical drum operating on an elasto-plastic and/or visco-plastic medium.
– Heavy cylindrical drums that are towed over hard, visco-plastic and/or sandy surfaces.
– Heavy, self-powered, cylindrical drums that drive themselves over hard, visco-plastic and/or sandy surfaces.
– Heavy, moving, cylindrical drums that brake themselves over hard, visco-plastic and/or sandy surfaces.

The cylindrical drum used in the modelling process can be thought of as the drum of an earthworks compactor or of an old fashioned steam-roller. Alternately it may be the steel tire or rigid wheel of a truck or trailer. While the use of rigid wheels may seem somewhat artificial to many, the assumption of rigidity makes the problem more mathematically tractable and gives experimentally testable results. The analyses presented here are generally two-dimensional only and the end effects of the drum and other three-dimensional effects have been typically ignored in the interests of mathematical tractability.

The chapter also continues the ideas developed in Chapter 1, namely that of the use of metrics to characterise a terrain and a particular load state. In this chapter, the ideas have been systematised by use of a matrix of terrain-wheel system constants.

REFERENCES

1. Al-Hussaini, M.M. & Gilbert, P.A. (1975). On the Stress Distribution Beneath a Circular Rigid Wheel. *Proc. 5th Int. Conf. ISTVS, Vol. 2, Michigan, U.S.A.*, pp. 335–365.
2. Ito, N. (1975). Theoretical Analysis of the Forces Acting about a Wheel Including Slip Sinkage. *Proc. 5th Int. Conf. ISTVS, Vol. 2, Michigan, U.S.A.*, pp. 311–333.
3. Wong, J.Y. & Reece, A.R. (1966). Soil Failure beneath Rigid Wheels. *Proc. 2nd Int. Conf., ISTVS, Quebec, Canada*, pp. 425–445.
4. Yong, R.N. & Windish, E.J. (1970). Determination of Wheel Contact Stresses for Measured Instantaneous Soil Deformation. *J. of Terramechanics, Vol. 7, No. 3/4*, pp. 57–67.
5. Yong, R.N. & Fattah, E.A. (1976). Prediction of Wheel Soil Interaction and Performance using the Finite Element Method. *J. of Terramechanics, 13, 4*, pp. 227–240.

6. Harrison, W.L. (1975). Shallow Snow Performance of Wheeled Vehicles. *Proc. 5th Int. Conf. ISTVS, Vol. 2, Michigan, U.S.A.*, pp. 589–614.
7. Akai, K. (1986). *Soil Mechanics.* pp. 169–192, Asakura Press, (In Japanese).
8. Terzaghi, K. (1943). *Theoretical Soil Mechanics.* pp. 118–143. John Wiley & Sons.
9. Terzaghi, K. (1943). *Theoretical Soil Mechanics.* pp. 367–415. John Wiley & Sons.
10. Janosi, Z. & Hanamoto, B. (1961). The Analytical Determination of Drawbar Pull as a Function of Slip for Tracked Vehicle. *Proc. 1st Int. Conf. on Terrain-Vehicle Systems*, Torio.
11. Bekker, M.G. (1961). *Off-the-Road Locomotion.* pp. 58–66. The University of Michigan Press.
12. Kacigin, V.V. & Guskov, V.V. (1968). The Basis of Tractor Performance Theory. *J. of Terramechanics, 5, 3*, pp. 43–66.
13. Terzaghi, K. & Peck, R.B. (1948). *Soil Mechanics in Engineering Practice.* pp. 167–175. John Wiley & Sons.
14. Wong, J.Y. & Reece, A.R. (1967). Prediction of rigid wheel performance based on the analysis of soil-wheel stresses – Part I. Performance of driven rigid wheels. *J. of Terramechanics, 4, 1*, pp. 81–98.
15. Wong, J.Y. (1967). Behaviour of soil beneath rigid wheels. *J. of Agric. Engng. Res., 12, 4*, pp. 257–269.
16. Yong, R.N. & Fattah, E.A. (1976). Prediction of wheel-soil interaction and performance using the finite element method. *J. of Terramechanics, 13, 4*, pp. 227–240.
17. Yong, R.N. & Fattah, E.A. (1975). Influence of Contact Characteristics on Energy Transfer and Wheel Performance on Soft Soil. *Proc. 5th Int. Conf. ISTVS, Vol. 2, Detroit, U.S.A.*, pp. 291–310.
18. Mogami, T. (1969). *Soil Mechanics.* pp. 479–622, Gihoudou Press, (In Japanese).
19. Wong, J.Y. & Reece, A.R. (1967). Prediction of Rigid Wheel Performance Based on the Analysis of Soil-Wheel Stresses – Part II. Performance of Towed Rigid Wheels. *J. of Terramechanics, 4, 2*, pp. 7–25.
20. Scott, R.F. (1965). *Principles of Soil Mechanics.* pp. 398–472. Addison-Wesley Publishing Company.
21. Wong, J.Y. (1978). *Theory of Ground Vehicles.* pp. 55–121. John Wiley & Sons.
22. Onafeko, O. & Reece, A.R. (1967). Soil Stresses and Deformations beneath Rigid Wheels. *J. of Terramechanics, 4, 1*, pp. 59–80.
23. Hiroma, T. & Ohta, Y. (1987). A Measurement of Normal and Tangential Stress Distribution under Rigid Wheel. *Terramechanics, Vol. 7*, pp. 7–13. The Japanese Society for Terramechanics, (In Japanese).
24. Krick, G. (1969). Radial and Shear Stress Distribution under Rigid Wheels and Pneumatic Tires Operating on Yielding Soils with Consideration of Tire Deformation. *J. of Terramechanics, 6, 3*, pp. 73–98.
25. Ogaki, M. (1984). The Normal and Tangential Stress Distribution Acting on the Contact Surface of the Rigid Wheel. *Terramechanics, Vol. 4*, pp. 12–15. The Japanese Society for Terramechanics, (In Japanese).
26. Oida, A., Satoh, A., Ito, H. & Triratanasirichai, K. (1991). Three Dimensional Stress Distributions on a Tire–Sand Contact Surface. *J. of Terramechanics, 28, 4*, pp. 319–330.

EXERCISES

(1) A rigid wheel of radius of $R = 10$ cm is running in straight forward motion under driving action on a sandy terrain at a rotational speed $R\omega = 30$ cm/s and with a slip ratio of $i_d = 10\%$. Calculate the total rolling time t and the total moving loss j during the rolling distance of 10 m.

(2) A rigid wheel of radius of $R = 20$ cm is running during driving action on a soft ground at a rotational speed $R\omega = 50$ cm/s and with a slip ratio of $i_d = 20\%$. The contact length of the rigid wheel against the terrain is measured as 10 cm. Calculate the transit time

t_e of the wheel over an arbitrary point X of the ground and the amount of slippage j_s of the soil at the point X.

(3) A rigid wheel is running during driving action on a soft ground at a slip ratio of $i_d = 15\%$. The radius of the wheel R is 8 cm, the speed of rotation $R\omega$ is 40 cm/s and the entry angle θ_f is $\pi/6$ rad. Calculate the slip velocity V_s and the amount of slippage j_d at the bottom-dead-center of the wheel.

(4) A rigid wheel is running during driving action on a soft ground at a slip ratio of $i_d = 30\%$. The entry angle θ_f at the beginning of contact of the wheel to the ground is $\pi/6$ rad. Can you show that the direction of the resultant velocity vector of a soil particle located on the contact point A is coincident with the gradient of tangent to the rolling locus e.g. trochoid curve of the wheel at this point?

(5) A rigid wheel of radius of $R = 50$ cm is running during driving action on a soft ground at a slip ratio of $i_d = 25\%$. All the soil particles on the contact part of the wheel are always rotating around the instantaneous center I. Calculate the coordinates of the center I relative to the origin of the wheel axle O.

(6) A rigid wheel of radius of $R = 20$ cm is running straight forward during braking action on a sandy terrain at a rotational speed of $R\omega = 50$ cm/s and at a skid of $i_b = -20\%$. Calculate the total rolling time and the total moving loss during the rolling distance of 10 m.

(7) A rigid wheel of radius of $R = 10$ cm is running during braking action on a soft ground at a rotational speed of $R\omega = 30$ cm/s and with skid of $i_b = -15\%$. The contact length of the rigid wheel to the terrain is measured as 10 cm. Calculate the transit time t_e of the wheel on an arbitrary point X of the ground and the amount of slippage j_s of the soil at the point X.

(8) A rigid wheel is running during braking action on a soft ground at the skid $i_b = -20\%$. The radius of the wheel R is 10 cm, the speed of rotation $R\omega$ is 50 cm/s and the entry angle θ_f is $\pi/6$ rad. Calculate the slip velocity V_s and the amount of slippage j_b at the bottom-dead-center of the wheel.

(9) A rigid wheel is running during braking action on a soft ground at a skid of $i_b = -30\%$. The entry angle θ_f at the beginning of contact of the wheel to the ground is $\pi/6$ rad. Demonstrate that the direction of the resultant velocity vector of a soil particle located on the contact point A is coincident with the gradient of tangent to the rolling locus e.g. trochoid curve of the wheel at this point.

(10) A rigid wheel of radius of $R = 30$ cm is running during braking action on a soft ground at a skid of $i_b = -50\%$. All the soil particles on the contact part of the wheel are always rotating around the instantaneous center I. Calculate the coordinates of the center I from the origin of the wheel axle O.

Chapter 3

Flexible-Tire Wheel Systems

In 1888, the Scottish inventor John Dunlop submitted a patent for a pneumatic tire for a tricycle. Since that time, many different forms of pleasure and industrial vehicles with soft and fluid-pressure tires have been developed. In construction and mining, especially, many common forms of excavation and loading machines such as, wheel-loaders, backhoes, tractor shovels and skid-steer loaders come equipped with special types of heavy duty rubber tires. Also, many pieces of modern civil engineering and mining haulage equipment, such as rigid and articulated-frame dump-trucks, motor-graders and rubber tired roller-compactors, operate through multiple sets of on-the-road or off-the-road rubber tires. Because these air or fluid-filled rubber tires are deliberately designed to deform with axle load or wheel torque, they are formally referred to as 'flexibly tired wheels'.

The interaction problem between a flexible tire and a particular terrain is an engineering problem that has been studied for many decades. In particular, experiments and theory have been developed to define the relations that exist between axle load, torque and ground reaction and the contact pressure distribution of the tire taking into consideration the compression and shear deformation characteristics of the terrain. Additionally, the relationship between thrust or drag, land locomotion resistance, effective driving or braking force and slip ratio or skid have been extensively studied for various types of terrain materials during both wheel driving and/or braking action.

For hard terrains, there exists a body of research which principally concerns itself with the motion of revolution of a tire, the kinematic straight forward motion equations and the tire's cornering characteristics during driving and braking action. For instance, Komandi [1] studied the circumferential force of a tire acting on a concrete pavement. Muro [2] concluded that the frictional work that develops between the tire of a heavy dump truck and a terrain will depend on the longitudinal and lateral amount of slippage. He also studied the cornering characteristics of the off-the-road tire.

However, to properly analyse the trafficability problem of a tire running on a soft terrain, it is necessary to take into account variations in the shear strength and deformation properties of the terrain. The reason for this is that various compaction and remolding effects can occur during passage of a tired wheel. These effects do not occur with very hard terrains. For soft terrains, Yong et al. [3] developed a new analytical process, based around the finite element method (FEM), which potentially can analyse the interaction problems between a tire and a clayey terrain. They consider the effect of the flexibility of the carcass of the tire on the distribution of the contact pressure which acts on the peripheral surface of the tire. They also claim that the analytical results agree well with available experimental data for various slip ratios. For soft ground Fujimoto [4] has analysed theoretically some matters relating to driving torque, rolling resistance and rut depth for a tire running in driving mode on a clayey soil. Forde [5] has portrayed the relationship between effective tractive effort and driving torque on a tire by use of several contour lines that utilise the parameters of slip ratio, amount of slippage and vertical ground reaction. Forde demonstrated that a higher effective tractive effort could be developed for a clayey soil having higher plastic index rather than a lower plastic index even if the consistencies and the water contents of both the clayey soils had the same values.

In the following sections of this Chapter, some methods for predicting the distribution of contact pressure acting on the peripheral surface of a tire as well as the fundamental characteristics of trafficability and the cornering characteristics of a tire will be developed based upon the mechanical characteristics of the structure of a tire. Also some aspects of land locomotion mechanics that utilise the equations of the movement of a tire running on a hard or soft terrain will be discussed. Further, a mathematical simulation method will be described that allows calculation of the trafficability of a tire during driving and braking actions to be undertaken. The simulation method will be then used to develop several analytical results.

3.1   TIRE STRUCTURE

An off-the-road tire, of the kind typically used in the construction and mining industries, has three principal functions. The first function is to maintain trafficability of the tire when it is sustaining high axle loads and when it is operating over soft and/or muddy terrains.

The second function is the relief of impact loads on, or generated by, the tire as they occur on uneven rough terrain such as rock masses or in forests. The third function is to guarantee the tight propagation of the driving and braking torques and the cornering forces that occur during turning motions on a terrain. Additionally, any vehicle carrying a large amount of cargo needs a high payload tire. Also, any vehicle running on a soft terrain needs a wide base tire of low contact pressure and high buoyancy. Further, when an off-the-road tire is running over a rocky terrain – such as blasted stone – or over a stump, a high resistance to abrasive wear or cut development is typically required of the tire. Moreover, where an off-the-road tire is required to run at high speed for a long period, a tire having a high heat resistance is demanded.

As shown in Figure 3.1, the structure of an off-the-road tire consists of various parts: namely the crown, shoulder, side wall and bead. The main surrounding part of the tire is called the 'carcass' or 'casing'. This part is required to sustain, a typically high, air pressure. The carcass has an important role to play in the development of effective driving or braking force in a tired wheel system by propagating the axle load, and the driving and braking torques to the terrain. It is also required to pass the ground thrust-reaction, drag, land locomotion resistance, and cornering forces through to the wheel.

Usually, a carcass is produced in several discrete stages. Firstly, rubber coated sheets are made through a gluing process in which 'cloths' made of cross folded tire cords (such as cotton, nylon or polyester) are developed and in which the voids and interstices are filled with rubber material. Then, a number of these rubber coated sheets are stuck together alternately with rubber material sandwiched between them. Bias tires belong to a group of tires in which the direction of the rubber coated sheets is biased to the central line of tread. On the other hand, radial tire belongs to another group of tire wherein the rubber coated sheets are directed radially i.e. at right angle to the central line of tread, and the carcass layers are bound together by several belts.

On the crown part of the tire, a thick rubber tread is used to protect the carcass. The rubber materials that comprise the tread must have a high resistance to abrasive wear, to cut

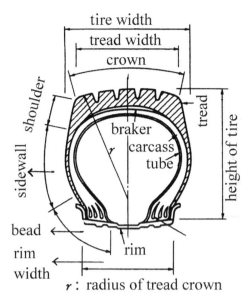

Figure 3.1. Structure of tire.

development and to the large amounts of heat that builds-up when a large torque is applied between the tire and terrain and when large amounts of slippage occur. For the various types of demand situation that occur in relation to the operation of off-road tires, several different configurations of tread pattern comprised of thick rubber plates and deep grooves have been developed.

On the outer part of the tire, a number of nylon cords or steel wires are used to reinforce and compensate for the large differences in rigidity that occur between crown and carcass. To protect the carcass, nylon cord is commonly used as a cushion material for bias tires and likewise steel wire is inserted in a part of the belt for radial tires. On the part of carcass which forms the skeleton of the tire, several sheets of nylon carcass are stuck to each other for bias tires whilst a steel carcass is used for radial tires.

On the sidewall part of the tire, special rubber materials which have a high bending resistance and a high weathering resistance are used to protect the deformation of the carcass. On the bead part of the tire, several ring shaped bead wires which are made of steel are used to tightly fix the tire carcass to the rim against the action of the high internal air pressure.

In the most common arrangement, the ply layers of the carcass are folded and wound around the bead wire.

To characterise the size and strength of an off-the-road tire, a designation of width of tire/flatness-rim diameter-ply rating (PR) like 45/65-45-50PR is used. The measuring units used in the tire industry for width of tire and rim diameter are inches. The flatness ratio (i.e. the tire aspect ratio) is defined as the ratio of height of the tire to the width of tire expressed as a percentage.

Figure 3.2 shows several representative tread patterns for various types of off-the-road tire. Sketch (a) shows a rib type tread which has several grooves that run parallel to the longitudinal direction of the tire. This configuration produces a high resistance against

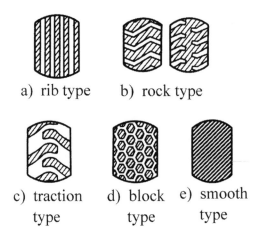

a) rib type   b) rock type

c) traction   d) block   e) smooth
   type   type   type

Figure 3.2. Tread patterns.

lateral slippage which is a property that gives a superior stability of operation to the tire. This tread pattern is often used for the front tires of motor graders because of the small rolling resistance and good lateral stability. Sketch (b) shows a rock type tread which is typically used for the tires of heavy dump trucks or shovel loaders which may be required to operating at quarry sites. The tread rubber in this case is of a kind that has a high resistance to abrasive wear and to the development of cuts. Sketch (c) illustrates a traction type tread which is designed to tightly transfer or couple driving and braking torques to the terrain. In this type, the direction of the groove is inclined to the central axis of the tread. This tread configuration is considered to be most effective for a driven tire when it is set up in a direction such that it discharges the soil, while for a towed tire it is best used set up in the reverse direction. Sketch (d) illustrates a block type tread which is massed with buttons and which is designed to have a high floatation. As contact area increases with increasing values of axle load, the bearing capacity e.g. the high buoyancy on a soft terrain, is maintained. In contrast, sketch (e) illustrates a smooth type tread which has no grooves on the peripheral surface of the tire. This type of tread pattern is used for the tires of compaction machinery, such as rubber tired rollers, which are required to develop high levels of compaction of earthfill.

3.2   STATIC MECHANICAL CHARACTERISTICS

When a (off-the-road) tire of radius R stands on a steel plate while sustaining an axle load W, as shown in Figure 3.3, the contact area A can be expressed by Eq. (3.1). This equation, however, idealises reality to a degree, in that it assumes that the tire deforms only at the contact interface with the terrain and that the contact pressure equals the tire air pressure p.

$$a = \sqrt{2Rf}$$

$$b = \sqrt{2rf}$$

$$A = \pi ab = 2\pi f \sqrt{Rr} \qquad (3.1)$$

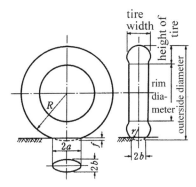

Figure 3.3. Deformation of contact part of tire on rigid plate.

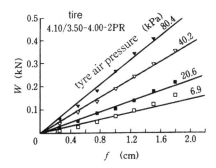

Figure 3.4. Relationship between static axle load W and amount of deformation f of buff tire [6].

and hence

$$W = 2\pi f \sqrt{Rr}\, p \qquad\qquad (3.2)$$

where r is the radius of the tread crown, f is the amount of deformation of the tire, and a and b are the major and minor axes of an elliptical contact area. From Eq. (3.2), it can be seen that the sustainable axle load W for a given tire pressure, is proportional to the tire deformation f.

To confirm or deny this relation, Figure 3.4 shows some experimental test results which Yong et al. [6] obtained for a treadless buff tire of 4.10/3.50-4.00-2PR in an study they undertook to discover the relation that exists between W and f. Their test results confirm, to a very good degree, the above relation Eq. (3.2) for various tire air pressures p.

Unfortunately, however, this simple idealised model is not valid for actual off-the-road tires because these are equipped with several types and thicknesses of tread rubber.

In the off-the-road tire situation, the amount of deformation of the tire f is made up of a deformation of the tread rubber, a deformation of the carcass of the tire and a compressive deformation of the contained air. For this more complex situation, the axle load W can be expressed as a non-linear relation, such as Eq. (3.3) which involves the total amount of deformation f.

$$W = \frac{f^2}{C_1 + C_2\{f/(p+p_0)\}} \qquad\qquad (3.3)$$

In this equation, C_1 is a constant which depends on the radius of the tire R, on the radius of the tread crown r, on the amount of deformation of the tire f, on the thickness of the tread, on the elastic modulus of the tread rubber, on the coefficient of fillness of the tread area in relation to the total contact area, on the coefficient of contact pressure distribution and on the stiffness of the tread rubber against lateral deformation. The parameter C_2 is a coefficient which can be determined from data relating to the radius of the tire R, to the radius of the tread crown r, and to the ratio of the volume of the elliptical segment at the contact part to the inside volume of the tire. The factor p is the tire air pressure and p_0 is defined as the ratio of the deformation work of the carcass at zero air pressure to the amount of variation of the inside volume of the tire.

Yong et al. [7] studied analytically the effects of elastic modulus and shear modulus of rigidity of a tire assuming a two-dimensional situation where two elastic cylinders are rolling. The cylinders have a contact length of $2a$ and sustain an axle load of P. The results of their study are contained in the following set of expressions.

$$\frac{1}{R_r} + \frac{1}{R_s} = \frac{4P}{\pi a^2}\left(\frac{1-v_s^2}{E_T} + \frac{1-v_T^2}{E_s}\right)\bigg/\left(1+\frac{k_s^2}{k_T^2}\right)$$

$$k_T = \frac{2}{\pi}\left(\frac{1-v_T}{G_T} + \frac{1-v_s}{G_s}\right)$$

$$k_s = \frac{1-2v_s}{G_T} - \frac{1-2v_s}{G_s}$$

$$G_T = \frac{E_T}{2(1+v_T)}$$

$$G_s = \frac{E_s}{2(1+v_s)} \tag{3.4}$$

For the particulars of this analytical case, R_T is the radius of the tire and R_s has a value of ∞. E_T is the elastic modulus of the tire, E_s is the elastic modulus of the terrain, G_T is the shear modulus of rigidity of the tire, G_s is the shear modulus of rigidity of the terrain and v_T and v_S are the Poisson's ratio of the tire and the terrain respectively. The value of the parameter v_t can be taken to have a value of 0.5 for the negligibly small variation in the volume of the tire that applies in this situation. The values of E_T and G_T of the tire can be calculated subsequent to measuring the values of E_s, v_s and a for a particular terrain.

Abeels et al. [8] developed an experimental apparatus as shown in Figure 3.5 to investigate the mechanical properties of tires. When a test tire was placed in a soil bin, which was filled with various kinds of soil material, and a test axle load W and torque Q was applied, the resulting contact length $2a$, the height of tire H_w and the width of tire B_w could be observed. Using these observations the researchers calculated a ratio of compression t_h and a ratio of flatness t_B of the tire using the following equations:

$$t_h = \frac{H_0 - H_w}{H_0} \times 100 \quad (\%) \tag{3.5}$$

$$t_B = \frac{B_0 - B_w}{B_0} \times 100 \quad (\%) \tag{3.6}$$

Here, H_0 and B_0 are the initial height and the initial width of the tire, respectively.

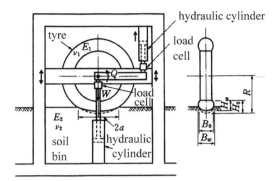

Figure 3.5. Measuring apparatus for elastic modulus of tire [8].

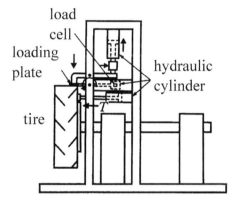

Figure 3.6. Test apparatus for deformation of tread and lateral rigidity of tire [8].

The same research group also developed another experimental apparatus – as shown in Figure 3.6. Through the use of this set-up they were able to investigate the lateral rigidity of tires and the characteristics of deformation of the tire's tread due to lateral load.

In this case, the lateral shear modulus of rigidity of a tire can be obtained from the shear strain calculated for the measured amount of deformation of the loaded point to which a cornering force T is applied.

The elastic modulus of the tread can be obtained by measuring the normal amount of deformation of the tread when a normally directed load is applied. In this specific piece of apparatus, the normal force is generated by use of a hydraulic jack via a lever which is pivoted on the upper frame. Under these same circumstances, the shear modulus of rigidity of the tread can also be obtained by measuring the amount of deformation of the loaded point when both a normal and a tangential force are applied simultaneously. These twin forces are generated by use of two hydraulic jacks via two levers pivoted on the upper and side frame respectively.

Yong et al. [7] determined that the elastic modulus E_T of tires equipped with several types of grooved tread rubber increases linearly with increasing tire air pressure p for constant axle load. Typical data to illustrate these results is given in Figure 3.7.

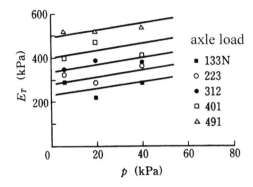

Figure 3.7. Relationship between modulus of elasticity E_T of tire and air pressure p [7] (tread tire 4.10/3.50-4.00-2pk).

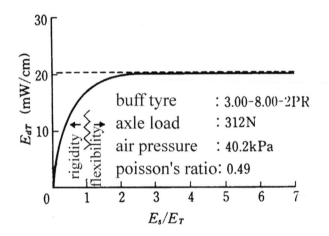

Figure 3.8. Relationship between deformation energy E_{dT} of tire and relative rigidity of terrain to tire E_s/E_T.

Yong et al. also used the finite element method, to demonstrate that the deformation energy E_{dT} of a tire varies with the ratio of the elastic modulus of the terrain E_s to the elastic modulus of the tire E_T for a constant axle load and tire air pressure [6]. Their results are shown in Figure 3.8. In this case, $E_s/E_T = 1.0$ can be considered to be a boundary in judging whether a tire is rigid or flexible. Thus, a tire may be deemed to act as a rigid wheel when E_s is less than E_T.

Fujimoto [9] determined that a critical tire air pressure p could be developed as the difference between the average contact pressure q under an axle load W and the rigid contact pressure of the carcass p_c which exists when a tire is self-standing by virtue of its own rigidity and when there is no applied axle load. The critical tire air pressure is the tire air pressure at which a tire changes from a rigid tire into a flexible tire.

A tire can be considered to be a rigid tire when the relation of Eq. (3.7) applies:

$$p \geq q - p_c \qquad (3.7)$$

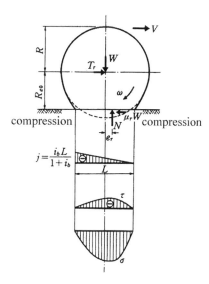

Figure 3.9. Ground reaction during pure rolling state of flexible tired wheel on hard terrain.

In these studies, it is very important to standardize test methods for the static mechanical characteristics of tires such as the elastic modulus and the shear modulus of rigidity. It is especially necessary to clearly define the mutual relationships between the tire air pressure, the elastic modulus and the shear modulus of rigidity of the tire.

3.3   DYNAMIC MECHANICAL PROPERTIES

3.3.1   Hard terrain

When a flexible tired wheel is running upon a hard terrain – such as an asphalt or concrete pavement, the amounts of deformation of the tread rubber and the carcass of the tire become very large, while the amounts of deformation in the terrain are negligibly small.

Suppose that we now analyse the force balances that apply when a tire is running on a hard terrain at a constant speed V with a constant rotation speed $R\omega$ and that we undertake this analysis alternately for the pure rolling state and for the driving and braking states.

In this situation we can note that for the pure rolling state, some compressive deformation will occur symmetrically on both sides of the contact part of the tire due to the axle load W. Also, as shown in Figure 3.9, the amount of slippage j at the end of the contact part of a tire of length L during braking action can be given as $i_b L/(1 + i_b)$ for a skid value i_b (<0). As well, the distributions of contact pressures e.g. the normal stress σ and shear resistance τ, deviate a little from their rest positions in the moving direction of the tire.

In the externally propelled rolling wheel case, the vertical ground reaction N must equal the axle load W and the effective braking force T_r acting in the moving direction of the wheel must equal the horizontal ground reaction $\mu_r W$ i.e. the rolling resistance. The direction of the resultant force comprised of N and $\mu_r W$ goes through the central axis of the wheel. The position of application of the resultant force can be expressed in terms of a vertical effective

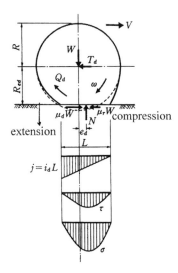

Figure 3.10. Ground reaction during driving action of flexible tired wheel on hard terrain.

rolling radius R_{e0} and a horizontal amount of eccentricity e_r. Using these parameters, the moment balance around the axle can be expressed as:

$$Ne_r = \mu_r W R_{e0} \tag{3.8}$$

and hence

$$e_r = \mu_r R_{e0} \tag{3.9}$$

where μ_r is the coefficient of rolling friction.

For a flexibly tired wheel operating in a driving state, the forward part of the carcass to the moving direction will expand due to compression effects whilst the rear part of the carcass will be extended due to tension effects upon the application of an axle load W and a driving torque Q_d. As shown in Figure 3.10, the amount of slippage j at the end of the contact part of the tire of the length L during driving action can be given as $i_d L$ for a slip ratio i_d (>0). In this case, the distributions of contact pressures e.g. normal stress σ and shear resistance τ deviate towards the moving direction of the tire. From considerations of force equilibrium, the vertical ground reaction N must equal the axle load W whilst the effective driving force T_d must equal the horizontal ground reaction $(\mu_d - \mu_r)W$ – where μ_d is the coefficient of driving resistance.

The position of application of the resultant ground reaction in this case can be expressed in terms of a vertical effective rolling radius R_{ed} and a horizontal amount of eccentricity e_d. Thence, from moment equilibrium around the axle we have:

$$Q_d = \mu_d W R_{ed}$$
$$Q_d - Ne_d - (\mu_d - \mu_r)W R_{ed} = 0 \tag{3.10}$$

and hence

$$e_d = \mu_r R_{ed} \tag{3.11}$$

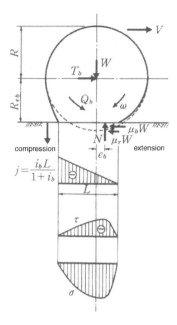

Figure 3.11. Ground reaction during braking action of flexible tired wheel on hard terrain.

In the initial stage of application of a small driving torque Q_d, $\mu_d < \mu_r$ the direction of action of the effective driving force T_d is coincident with the moving direction of the wheel. When the driving torque Q_d increases more and $\mu_d = \mu_r$, T_d becomes zero and the wheel will run in a self-propelling state. For a large driving torque Q_d, the effective driving force T_d applies oppositely to the moving direction of the wheel as shown in the diagram.

Harlow et al. [10] determined that all the shear resistance vectors acting on the tread surface were located in the moving direction of the wheel and along the longitudinal center line of the contact part of the tire.

For a flexibly tired wheel operating in braking mode, the forward part of the carcass to the moving direction will be extended due to tension effects and the rear part of the carcass will be expanded due to compression effects through the application of an axle load W and a braking torque Q_b. As shown in Figure 3.11, the amount of slippage j at the end of the contact part of the tire of length L during braking action may be given as $i_b L/(1 + i_b)$ for a skid value $i_b(<0)$. In this situation the distributions of contact pressures e.g. the normal stress σ and shear resistance τ progress towards the moving direction of the tire. Also, the vertical ground reaction N equals the axle load W and the effective braking force T_b equals the horizontal ground reaction $(\mu_b + \mu_r)W$ where μ_b is the coefficient of braking resistance. In a similar manner to the above cases, the position of application of the resultant ground reaction can be expressed in terms of a vertical effective rolling radius R_{eb} and a horizontal amount of eccentricity e_b. Thence, moments around the axle can be equated to yield the following equation.

$$Q_b = \mu_b W R_{eb}$$

$$-Q_b - N e_b + (\mu_b + \mu_r) W R_{eb} = 0 \qquad (3.12)$$

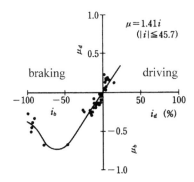

Figure 3.12. Measured relationship between longitudinal component of friction μ_d and slip ratio i_d of heavy dump truck [11].

therefore

$$e_b = \mu_r R_{eb} \tag{3.13}$$

The relationships that exist between the coefficients of driving and braking resistance μ_d, μ_b, and the slip ratio i_d and the skid i_b for any particular off-the-road tire depend on the shape of the groove, the material and the surface roughness of the tread rubber. For example, Figure 3.12 shows the μ–i curve for an off-the-road tire for a heavy dump-truck. The data was measured in-situ while the tire was running on a hard terrain of decomposed weathered granite based sandy soil [11]. The tire was of a 1.00-35-36PR type.

3.3.2   Soft terrain

Relative to the problem of determining the trafficability of an off-the-road tire that is running on a soft terrain, Turnage [12] proposed that the coefficient of traction and the tractive efficiency of a tire could be estimated experimentally from the next wheel mobility numbers N_s and N_c as developed in the following equations:

For sandy terrain:

$$N_s = \frac{G(BD)^{\frac{3}{2}}}{W} \cdot \frac{\delta}{H} \tag{3.14}$$

For clayey terrain:

$$N_c = \frac{CBD}{W} \left(\frac{\delta}{H} \right)^{\frac{1}{2}} \cdot \frac{1}{1 + B/2D} \tag{3.15}$$

Here, C is the cone index of the terrain, G is the gradient of the cone index e.g. the cone index divided by the depth of penetration, W is the axle load of the tire, B is the width of the wheel, D is the diameter of the wheel, H is the height of the tire, and δ is the amount of tire deformation.

Figure 3.13(a) shows the relationship between the coefficient of traction μ_{20}, the tractive efficiency η_{20} at slip ratio $i_d = 20\%$ and the wheel mobility number N_s for a sandy terrain.

Likewise, Figure 3.13(b) shows the relationship between μ_{20}, η_{20} at $i_d = 20\%$ and wheel mobility number N_c for a clayey terrain.

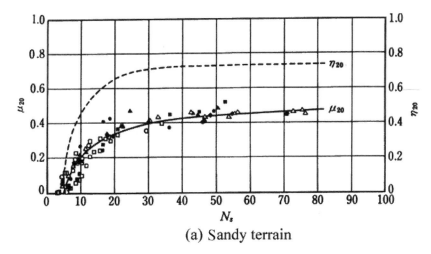

(a) Sandy terrain

Figure 3.13a. Relationship between coefficient of traction μ_{20}, tractive efficiency η_{20} and wheel mobility number N_s [12] for a sandy terrain.

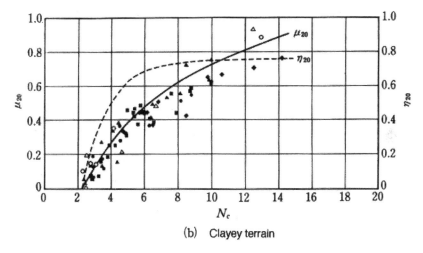

(b) Clayey terrain

Figure 3.13b. Relationship between coefficient of traction μ_{20}, tractive efficiency η_{20} and wheel mobility number, N_c [12] for a clayey terrain.

Yong et al. [3] investigated the effects of the tread pattern of a tire on the effective driving force during driving action for various levels of tire air pressure. As a result, it was determined that a traction type tread can improve, quite remarkably, the effective driving force, whilst a more rigid tire having high air pressure can develop the most effective driving force for a sandy terrain. For a clayey terrain a more flexible tire having low air pressure can develop the most effective driving force. For example, Figure 3.14 shows a relationship that prevails for various values of energy E_{dt} and slip ratio i_d for an off-the-road tire with mounted tread rubber. The tire is inflated to an air pressure of 40 kPa such as would be indicated for a sandy terrain. From this diagram, it is evident that the slippage energy

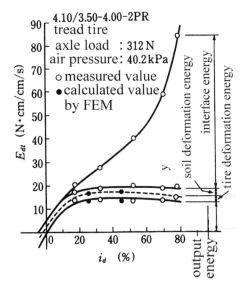

Figure 3.14. Relationship between energy E_{dt} and slip ratio i_d of tread tire [3].

between the tire and the terrain increases parabolically with increasing values of i_d and that the output energy takes a maximum value at some value of slip ratio.

In general, the shape of the distribution of contact pressure of a tire running on a soft terrain is essentially identical to that of a rigid wheel moving in pure rolling, driving or braking states as has been already discussed in the previous Chapter. Further, Krick [13] and Oida et al. [14] confirmed that contact pressure tends to decrease relatively with increasing deformation of the contact part of the tire.

Söhne [15] investigated the effect of the stress distribution under a tire on the deformation of the terrain and on the amount of compaction of the substrate soil. In these studies, it is necessary to vary the type of tire, the axle load, the thickness of lift and the number of compaction passes required to achieve an optimum compaction of soil as a consequence of the existence of a finite sized pressure bulb that occurs in the terrain under the tire. The size of this bulb varies with axle load and soil properties.

In these studies, it was found that a most effective method of compaction for increasing the dry density of a soil and its shear strength could be developed by creating an alternating shear stress in the soil. This effect could be practically obtained by developing a coupled land locomotion system comprised of a pure rolling front tire and a driven rear tire [16].

To analyse the specific mechanisms of soil compaction, Bolling [17] calculated the trochoidal rolling locus of a tire running at various slip ratios and thence measured the vertical distribution of the normal earth pressure from a plate shear test by use of a plate mounted with a given tread rubber which moved in the soil along the rolling locus.

(1) *During driving action*

When the tire air pressure p of a tire running on a soft terrain is larger than the difference between the average contact pressure q and the rigid contact pressure of the carcass p_C as described in Eq. (3.7), the tire can be treated as a 'rigid wheel' because the tire does

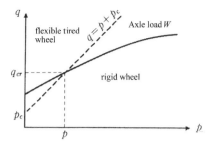

Figure 3.15. Relationship between tire air pressure p and average contact pressure q (p_C = rigid contact pressure).

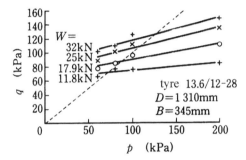

Figure 3.16. Relationship between tire air pressure p and average contact pressure for various axle loads W on tire [19].

not deform. Bekker [18] proposed that the critical average contact pressure q_{cr} which distinguishes between a 'rigid wheel' and a 'flexible wheel' equals the sum of the tire air pressure p and the rigid contact pressure of the carcass p_C on a hard terrain, as shown in Figure 3.15. Unfortunately though and in general, the rigid contact pressure of the carcass p_C on a soft terrain can not easily be determined since p_C depends not only on the tire air pressure and the weight of tire but also on the properties of the soil that comprises the soft terrain. The previous theory, as mentioned in Section 2.2, can be applied to provide an estimate for a rigid tire since the average contact pressure q, e.g. the axle load W divided by the contact area, is less than the critical average contact pressure q_{cr}.

To investigate these findings, Schwanghart [19] measured the contact area of an off-the-road tire 13.6/12-28 of diameter 1310 mm and of width 345 mm under various axle loads W when the tire system was operating on a loose accumulated sandy loam terrain of water content 15%. From these observations he developed a relationship between average contact pressure q and the tire air pressure p as shown in Figure 3.16. From this diagram, the tire can be seen to function as a rigid wheel in those regions where $q \leq p + p_C$. For constant terrain conditions and for a constant axle load W, the average contact pressure q increases with increments in tire air pressure p due to corresponding decrements in contact area.

Further, for constant terrain conditions and for a constant tire air pressure p, the average contact pressure q tends to increase with increments in axle load W, while the contact area also increases with increments in W.

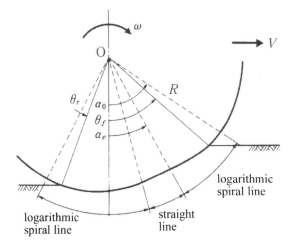

Figure 3.17. Flexible deformation of tire (straight logarithmic spiral line).

When the tire air pressure p has a comparatively low value and is less than the difference between the average contact pressure q and the rigid contact pressure of the carcass p_C, the tire can be treated as a 'flexible wheel' because the tire can deform.

There are many hypotheses about the shape of the flexible deformation of the contact part of the tire. Karafiath et al. [20] proposed that the shape of the deformation of the contact part of a tire could be considered to be comprised of a central straight part with logarithmic spiral line portions on both sides as shown in Figure 3.17. The radius r of the forward logarithmic spiral line can then be expressed as a function of an arbitrary central angle α, as follows:

$$r = R \exp\{\beta(\alpha - \alpha_0)\} \tag{3.16}$$

where R is the radius of tire, β is a constant value, α_0 is the central angle at the beginning of the logarithmic spiral line, and $\alpha_0 = \theta_f + \pi/36$ rad for the entry angle θ_f.

The angle α is taken to lie arbitrarily in the range $\alpha_e \leq \alpha \leq \alpha_0$ where α_e is the central angle at the beginning of the straight line. Using the same approach, the radius of the rear logarithmic spiral line can be likewise calculated.

Blaszkiewicz [21] developed a real-time apparatus for measuring the depth of the rut generated and the radial, longitudinal and lateral amounts of deformation of a tire running on a soft terrain. Using an assumed radial deformation function taken as a polynomial expression, he made a mathematical model which expressed the total shape of the deformation of the contact part of the tire provided that the shape of cross section of the contact part of the tire due to the longitudinal and lateral deformation could be approximated as an ellipse.

Wong [22] proposed a somewhat simpler deformation shape for a tire using circular arc segments as shown in Figure 3.18.

Relative to these conditions, let us now consider the overall equilibrium and force balance that exists between the axle load W, the driving torque Q_d, the effective driving force T_d during driving action, and the distributions of contact pressure $\sigma(\theta)$, $\tau(\theta)$ which act on the contact part of a flexible tired wheel. As shown in the diagram, the central angle of the straight part $\overline{AB}$ which is flattened due to the deformation of the tire is $2\theta_C$ and the central angles of the circular arcs $\hat{BC}$ and $\hat{AD}$ are $\theta_f - \theta_C$ and $\theta_r - \theta_C$ respectively. The original

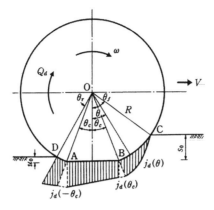

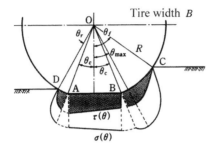

Figure 3.18. Distributions of amount of slippage $j_d(\theta)$ and ground reaction $a(\theta)$, $\tau(\theta)$ of contact part of tire (during driving action).

radius of the tire is designated as R and the central angle of the radius vector, measured counterclockwise from the vertical axis, is designated as θ.

The respective amounts of slippage $j = j_d(\theta)$ of the circular part $\widehat{BC}$, the straight part $\overline{AB}$ and the circular part $\widehat{AD}$ of the contact part of the tire may be expressed, using the previous Eq. (2.16), as ;

For $\theta_C \leqq \theta \leqq \theta_f$
$$j_d(\theta) = R\{(\theta_f - \theta) - (1 - i_d)(\sin \theta_f - \sin \theta)\} \qquad (3.17)$$

For $-\theta_C \leqq \theta < \theta_C$
$$j_d(\theta) = j_d(\theta_C) + R(\sin \theta_C - \sin \theta)i_d \qquad (3.18)$$

For $-\theta_r \leqq \theta < -\theta_C$
$$j_d(\theta) = j_d(-\theta_C) + R \int_\theta^{-\theta_C} \{1 - (1 - i_d)\cos \theta\}d\theta$$
$$= j_d(-\theta_C) - R\{(\theta_C + \theta) - (1 - i_d)(\sin \theta_C + \sin \theta)\} \qquad (3.19)$$

The distributions of contact pressure $\sigma(\theta)$ and $\tau(\theta)$ take positive values for the whole range of the contact area of the tire. The angle $\delta(\theta)$ between the resultant pressure $p(\theta)$ and the radial direction of the tire may be defined as $\delta(\theta) = \tan^{-1}\{\tau(\theta)/\sigma(\theta)\}$. The amount of sinkage s_0 of the bottom-dead-center just under the central axis of the tire and the amount of rebound u_0 just after the passing of the tire can be expressed as:

$$s_0 = R(\cos\theta_C - \cos\theta_f) \qquad (3.20)$$

$$u_0 = R(\cos\theta_C - \cos\theta_r) \qquad (3.21)$$

Further, the resultant pressure $p(\theta)$ can be expressed, by use of an elemental length of rolling locus $d(\theta)$ in the direction of the applied stress $q(\theta)$ as shown in previous Eq. (2.36), a coefficient of modification ξ to allow for dynamic effects, and an angle of inclination of the effective driving force $\zeta = \tan^{-1}(T_d/W\cos\beta)$, as follows:

For $\theta_{max} \leqq \theta \leqq \theta_f$

$$p(\theta) = \frac{k_1\xi\{d(\theta)\}^{n_1}}{\cos(\zeta + \theta - \delta)} \qquad (3.22)$$

For $\theta_C \leqq \theta < \theta_{max}$

$$p(\theta) = \frac{k_1\xi\{d(\theta_{max})\}^{n_1} - k_2\{d(\theta_{max}) - d(\theta)\}^{n_2}}{\cos(\zeta + \theta - \delta)} \qquad (3.23)$$

For $-\theta_C \leqq \theta < \theta_C$

$$p(\theta) = \frac{k_1\xi\{d(\theta_{max})\}^{n_1} - k_2\{d(\theta_{max}) - d(\theta)\}^{n_2}}{\cos(\xi + \theta - \delta)} \qquad (3.24)$$

For $-\theta_r \leqq \theta < -\theta_C$

$$p(\theta) = \frac{||k_1\xi\{d(\theta_{max})\}^{n_1} - k_2\{d(\theta_{max}) - d(\theta_C)\}^{n_2} - k_2[d(\theta_{max}) - \{d(\theta_C) + d'(\theta)\}]^{n_2}||}{\cos(\zeta + \theta - \delta)}$$

$$(3.25)$$

Here, θ_{max} is the central angle θ at which $p(\theta)$ takes a maximum value and $d'(\theta)$ is a function which integrates the previous Eq. (2.36) from $\theta = \theta$ to $\theta = -\theta_c$. In the above equations, the coefficients k_1, k_2 and the indices n_1, n_2 are a set of terrain-tire system constants which can be determined from standard plate loading and unloading tests.

Thence, the distribution of normal stress $\sigma(\theta)$ may be given by:

$$\sigma(\theta) = p(\theta)\cos\{\delta(\theta)\} \qquad (3.26)$$

For $-\theta_C \leqq \theta \leqq \theta_C$, $\sigma(\theta)$ can be written as follows – assuming that the applied resultant stress is constant e.g. $p(\theta) = p_g$.

$$\sigma(\theta) = p_g\cos\{\delta(\theta)\} \qquad (3.27)$$

An expression for the distribution of shear resistance $\tau(\theta)$ can be obtained by substituting the amount of slippage $j_d(\theta)$ – given in the previous Eqs. (3.17), (3.18) and (3.19) – into the previous Eqs. (2.8), (2.9). This action yields the following equation.

$$\tau(\theta) = \{c_a + \sigma(\theta)\tan\phi\}[1 - \exp\{-aj_d(\theta)\}] \qquad (3.28)$$

where c_a and ϕ are, respectively, the adhesion and the angle of friction between a tire and a terrain. These values comprise another set of terrain-tire system constants which can be determined from the tire segment plate traction test.

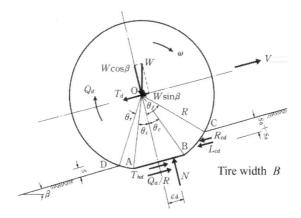

Figure 3.19. The various forces acting on a tire climbing up sloped terrain during driving action.

As shown in Figure 3.19, the axle load W, the driving torque Q_d and the apparent effective driving force T_{d0} of a rolling tire which is climbing up a sloping terrain of angle β during driving action, can be expressed by the following equation:

$$W \cos \beta = BR \left[\int_{\theta_C}^{\theta_f} \{\sigma(\theta) \cos \theta + \tau(\theta) \sin \theta\} \, d\theta + \int_{-\theta_C}^{\theta_C} p_g \cos\{\delta(\theta)\} \cos \theta \, d\theta \right.$$

$$\left. + \int_{-\theta_r}^{-\theta_C} \{\sigma(\theta) \cos \theta + \tau(\theta) \sin \theta\} \, d\theta \right] \tag{3.29}$$

$$Q_d = BR^2 \left\{ \int_{\theta_C}^{\theta_f} \tau(\theta) \, d\theta + \int_{-\theta_C}^{\theta_C} \tau(\theta) \cos \theta \, d\theta + \int_{-\theta_r}^{-\theta_C} \tau(\theta) \, d\theta \right\} \tag{3.30}$$

$$T_{d0} = BR \left[\int_{\theta_C}^{\theta_f} \{\tau(\theta) \cos \theta - \sigma(\theta) \sin \theta\} \, d\theta + \int_{-\theta_C}^{\theta_C} \tau(\theta) \cos \theta \, d\theta \right.$$

$$\left. + \int_{-\theta_r}^{-\theta_C} \{\tau(\theta) \cos \theta - \sigma(\theta) \sin \theta\} \, d\theta \right] \tag{3.31}$$

Additional to the above equation, the apparent effective driving force T_{d0} can be further expressed as the difference between the thrust T_{hd}, the compaction resistance R_{cd} and the slope resistance $W \sin \beta$ as follows:

$$T_{d0} = T_{hd} - R_{cd} - W \sin \beta \tag{3.32}$$

where

$$T_{hd} = BR \int_{-\theta_r}^{\theta_r} \tau(\theta) \cos \theta \, d\theta \tag{3.33}$$

$$R_{cd} = BR \left\{ \int_{\theta_C}^{\theta_f} \sigma(\theta) \sin \theta \, d\theta + \int_{-\theta_r}^{-\theta_C} \sigma(\theta) \sin \theta \, d\theta \right\} \tag{3.34}$$

In this case, the amount of slip sinkage is not considered. The thrust T_{hd} is set equal to the integral of $\tau(\theta) \cos \theta$ acting in the moving direction of the tire. The compaction resistance R_{cd} is the land locomotion resistance that occurs as a consequence of the formation of a rut of static depth s_0. In magnitude it is equal to the integral of $\sigma(\theta) \sin \theta$ and it acts in the opposite direction to the moving direction of the tire.

Further, the driving force can be set equal to Q_d/R. For flat terrains, the horizontal component $(Q_d/R)_h$ can be expressed as the sum of a number of components. These are the total compaction resistance L_{cd}, e.g. the land locomotion resistance calculated taking into consideration the amount of slip sinkage, and the actual effective driving force T_d, as shown in the previous Eq. (2.31).

The amount of slip sinkage s_s at an arbitrary point X on the terrain can be calculated from the amount of slippage of soil j_s given in the previous Eq. (2.53) as follows:

$$
s_s = c_0 \sum_{n=1}^{Nf} \{p(\theta_n) \cos(\theta_n - \delta_n)\}^{c_1} \left\{ \left(\frac{n}{N} j_s \right)^{c_2} - \left(\frac{n-1}{N} j_s \right)^{c_2} \right\}
$$

$$
+ c_0 \sum_{n=Nf}^{Nr} \{p(\theta_n) \cos \delta_n\}^{c_1} \left\{ \left(\frac{n}{N} j_s \right)^{c_2} - \left(\frac{n-1}{N} j_s \right)^{c_2} \right\}
$$

$$
+ c_0 \sum_{n=Nr}^{N} \{p(\theta_n) \cos(\theta_n - \delta_n)\}^{c_1} \left\{ \left(\frac{n}{N} j_s \right)^{c_2} - \left(\frac{n-1}{N} j_s \right)^{c_2} \right\} \tag{3.35}
$$

where

$$
\theta_n = \frac{n}{N}(\theta_f + \theta_r) \qquad \delta_n = \tan^{-1}\left\{ \frac{\tau(\theta_n)}{\sigma(\theta_n)} \right\}
$$

$$
N_f = N \frac{\theta_f - \theta_c}{\theta_f + \theta_c} \qquad N_r = N \frac{\theta_f + \theta_c}{\theta_f + \theta_r}
$$

In the above equation, the coefficient c_0, and the indices c_1, c_2 are the terrain-tire system constants obtained from the tire segment plate slip sinkage test.

The rut depth s associated with the passage of a tire can be calculated by use of the previous Eq. (2.55) taken with the values of the static amount of sinkage s_0 and u_0 given in the previous Eqs. (3.20), (3.21) and the amount of slip sinkage s_s given in the above equation. Thence, in a manner similar to that employed to generate the previous Eq. (2.56), the total amount of compaction L_{cd} can be calculated using the following equation:

$$
L_{cd} = k_1 \xi B \int_0^s z^{n_1} dz \tag{3.36}
$$

Next, the actual effective driving force T_d can be expressed as the difference between the tangential component of the driving force $(Q_d/R)_s$ in the direction of the surface of terrain, and the total compaction resistance L_{cd} (given in the above equation) and the slope resistance $W \sin \beta$ as follows:

$$
T_d = \left(\frac{Q_d}{R} \right)_s - L_{cd} - W \sin \beta \tag{3.37}
$$

In Figure 3.19, the product of the normal ground reaction N to the sloped terrain surface and the apparent amount of eccentricity e_{d0} can be seen to be equal to the integral of the elemental ground pressure distribution force couples taken over the range $\theta = -\theta_r$ to $\theta = \theta_f$. More particularly, this integral is comprised of the sum of:

(i) the product of the normal component of the resultant pressure to the sloped terrain surface $p(\theta) RB \cos\theta \cos(\theta - \delta) d\theta$ acting on an elemental area $RB \cos\theta \, d\theta$ at an arbitrary point on the contact part $\hat{BC}$ and $\hat{AD}$ to the terrain and with moment arm length $R \sin\theta$ and

(ii) the product of the normal component of the resultant pressure to the sloped terrain surface $p(\theta)RB \cos\theta_C \cos\delta \, d\theta$ acting on an elemental area $RB \cos\theta_C \, d\theta$ at an arbitrary point on the contact part $\overline{AB}$ and with a moment arm length $R \cos\theta_C$.

Thence, the apparent amount of eccentricity e_{d0}, without considering any slip sinkage, can be calculated as:

$$N = W \cos\beta \tag{3.38}$$

$$e_{d0} = \frac{BR^2}{W \cos\beta} \left\{ \int_{\theta_C}^{\theta_f} p(\theta) \cos\theta \cos(\theta - \delta) \sin\theta \, d\theta \right.$$

$$\left. + \cos^2\theta_c \int_{-\theta_C}^{\theta_C} p(\theta) \cos\delta \, d\theta + \int_{-\theta_r}^{-\theta_C} p(\theta) \cos\theta \cos(\theta - \delta) \sin\theta \, d\theta \right\} \tag{3.39}$$

However, to calculate the actual amount of eccentricity e_d, it is necessary to modify the above equation to allow for the effects of slip sinkage.

The effective input energy E_1 supplied by the driving torque Q_d acting on a tire may be equated to the sum of a number of output energies: namely a compaction energy component E_2 – associated with the development of a rut under the tire, a slippage energy component E_3 – associated with the development of shear deformation at the interface between tire and terrain, an effective driving force energy component E_4 e.g. the traction work required to draw another vehicle and a potential energy component E_5 that comes into play when the vehicle or wheel is on a slopped terrain. Since the system must be in energy balance the following equation applies:

$$E_1 = E_2 + E_3 + E_4 + E_5 \tag{3.40}$$

When the radius vector of the tire rotates from an entry angle θ_f at the beginning of the contact part of the tire to the terrain to the central angle $\theta_f + \theta_r$, the moving distance of the tire is equal to $R(\theta_f + \theta_r)(1 - i_d)$ during the rotation of an arbitrary point on the peripheral contact part of $R(\theta_f + \theta_r)$. The individual energy components that arise during the rotation of the tire can be computed as follows:

$$E_1 = Q_d(\theta_f + \theta_r) \tag{3.41}$$

$$E_2 = L_{cd}R(\theta_f + \theta_r)(1 - i_d) \tag{3.42}$$

$$E_3 = Q_d(\theta_f + \theta_r)i_d \tag{3.43}$$

$$E_4 = T_d R(\theta_f + \theta_r)(1 - i_d) \tag{3.44}$$

$$E_5 = WR(\theta_f + \theta_r)(1 - i_d)\sin \beta \qquad (3.45)$$

Further, the amount of energy produced or consumed per unit of time can be calculated from knowledge of the peripheral speed $R\omega$ of the wheel and the moving speed V of the system in the direction of an arbitrarily sloped terrain. For these calculations, the following equations can be used.

$$E_1 = Q_d\omega = \left(\frac{Q_d}{R}\right)_s \frac{V}{1 - i_d} \qquad (3.46)$$

$$E_2 = L_{cd}R\omega(1 - i_d) = L_{cd}V \qquad (3.47)$$

$$E_3 = Q_d\omega i_d = \left(\frac{Q_d}{R}\right)_s \frac{i_d V}{1 - i_d} \qquad (3.48)$$

$$E_4 = T_dR\omega(1 - i_d) = T_dV \qquad (3.49)$$

$$E_5 = WR\omega(1 - i_d)\sin \beta = WV\sin \beta \qquad (3.50)$$

Substituting the previous Eq. (3.37) into the above equations, the requisite energy equilibrium balance of Eq. (3.40) can be confirmed.

The optimum effective driving force T_{dopt} during driving action can be defined as the effective driving force T_d at the optimum slip ratio i_{dopt} when the effective driving force energy E_4 takes a maximum value for a constant circumferential speed $R\omega$. Additionally, the tractive efficiency E_d can be calculated from the previous Eq. (2.68).

(2) During braking action

Similar to the driving action case, a simplified shape of deformation of a tire can be adopted as shown in Figure 3.20.

Here the central angle of the straight line section of the flat part $\overline{AB}$ is $2\theta_C$, and the central angles of the arc portions $\hat{BC}$ and $\hat{AD}$ are $\theta_f - \theta_C$ and $\theta_r - \theta_C$ respectively. The angle θ is the central angle of the radius vector and is defined to be measured counter clockwise from the vertical line of the tire. The parameter R is the original radius of the tire and B is the width of the tire.

Let us now consider the force balance that exists between the axle load W, the braking torque Q_b and the effective braking force T_b during braking action and the distributions of the contact pressures $\sigma(\theta)$ and $\tau(\theta)$ that must act on the contact parts of the tire and the terrain.

The amount of slippage $j = j_b(\theta)$ on the respective contact parts of the arc $\hat{BC}$, the straight line segment $\overline{AB}$ and the arc $\hat{AD}$ can be determined by use of the previous Eq. (2.71), i.e.

For $\theta_C \leqq \theta \leqq \theta_f$

$$j_b(\theta) = R\left\{(\theta_f - \theta) - \frac{1}{1 + i_b}(\sin \theta_f - \sin \theta)\right\} \qquad (3.51)$$

For $-\theta_C \leqq \theta \leqq \theta_C$

$$j_b(\theta) = j_b(\theta_C) + R(\sin \theta_C - \sin \theta)\frac{i_b}{1 + i_b} \qquad (3.52)$$

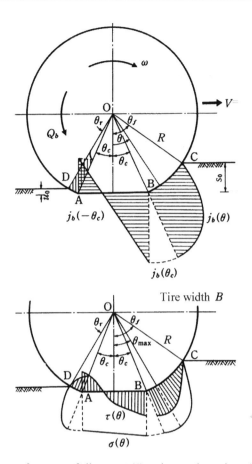

Figure 3.20. Distributions of amount of slippage $j_b(\theta)$ and ground reaction $a(\theta)$, $\tau(\theta)$ of contact part of tire (during braking action).

For $-\theta_r \leqq \theta \leqq -\theta_C$

$$j_b(\theta) = j_b(-\theta_c) + R \int_\theta^{-\theta_C} \left(1 - \frac{1}{1+i_b}\cos\theta\right) d\theta$$

$$= j_b(-\theta_c) - R\left\{(\theta_C + \theta) - \frac{1}{1+i_b}(\sin\theta_C \sin\theta)\right\} \tag{3.53}$$

For the central angle $\theta = \alpha$ at the maximum value $j_b(\theta) = j_p$, the shear resistance $\tau(\theta)$ takes a positive value for $0 \leq j_b(\theta) \leq j_p$ in the range of $\theta_f \geq \theta \geq \alpha$ and takes a negative value for $j_b(\theta) < j_q$ in the range of $\alpha \geq \theta \geq -\theta_r$ as mentioned previously in Eq. (2.89).

On the other hand, the normal stress $\sigma(\theta)$ takes a positive value for the whole range of the contact area. The angle δ between the applied resultant stress $p(\theta)$ and the radial direction of the contact part of tire is determined as $\delta = \tan^{-1}\{\tau(\theta)/\sigma(\theta)\}$.

The magnitude of the sinkage s_0 just under the central axis of the tire and the amount of rebound u_0 of the terrain just after the passage of the tire, respectively, are given in the previous Eqs. (3.20) and (3.21).

The applied resultant stress $p(\theta)$ can be determined given an elemental length of the component of the rolling locus $d(\theta)$ – as obtained from previous Eq. (2.85) – in the direction of the applied stress $q(\theta)$, the coefficient of modification ξ considering the dynamic mechanics of the tire, and the angle of the effective braking force $\zeta = \tan^{-1}(T_b/W\cos\beta)$ as in the following equations:

For $\theta_{max} \leqq \theta \leqq \theta_f$

$$p(\theta) = \frac{k_1\xi\{d(\theta)\}^{n_1}}{\cos(\varsigma - \theta + \delta)} \tag{3.54}$$

For $\theta_C \leqq \theta < \theta_{max}$

$$p(\theta) = \frac{k_1\xi\{d(\theta_{max})\}^{n_1} - k_2\{d(\theta_{max}) - d(\theta)\}^{n_2}}{\cos(\varsigma - \theta + \delta)} \tag{3.55}$$

For $-\theta_C \leqq \theta < \theta_C$

$$p(\theta) = \frac{k_1\xi\{d(\theta_{max})\}^{n_1} - k_2\{d(\theta_{max}) - d(\theta_C)\}^{n_2}}{\cos(\zeta - \theta + \delta)} \tag{3.56}$$

For $-\theta_r \leqq \theta < -\theta_C$

$$p(\theta) = \frac{||k_1\xi\{d(\theta_{max})\}^{n_1} - k_2\{d(\theta_{max}) - d(\theta_C)\}^{n_2} - k_2[d(\theta_{max}) - \{d(\theta_C) + d'(\theta)\}]^{n_2}||}{\cos(\zeta - \theta + \delta)}$$

$$\tag{3.57}$$

where θ_{max} is the central angle when $p(\theta)$ takes a maximum value, and $d'(\theta)$ is the length of the component of the rolling locus integrating the previous Eq. (2.85) from $\theta = \theta$ to $\theta = \theta_C$. Thence, the distribution of normal stress $\sigma(\theta)$ can be expressed as:

$$\sigma(\theta) = p(\theta)\cos\{\delta(\theta)\} \tag{3.58}$$

Assuming that the applied resultant stress is constant, e.g. $p(\theta) = p_g$, then $\sigma(\theta)$ is given as:

$$\sigma(\theta) = p_g\cos\{\delta(\theta)\} \tag{3.59}$$

Similarly, the distribution of shear resistance $\tau(\theta)$ can be calculated as discussed in relation to the previous Eq. (2.89). By substituting the amount of slippage $j_b(\theta)$ given in Eqs. (3.51), (3.52) and (3.53) into the previous Eqs. (2.8) and (2.9) the following results are obtained.

For the traction state: $0 \leqq j_b(\theta) \leqq j_p$

$$\tau(\theta) = \{c_a + \sigma(\theta)\tan\phi\}[1 - \exp\{-aj_b(\theta)\}]$$

For the untraction state: $j_q < j_b(\theta) < j_p$

$$\tau(\theta) = -\{c_a + \sigma(\theta)\tan\phi\}[1 - \exp\{-a[j_q - j_b(\theta)]\}] \tag{3.60}$$

For the reciprocal traction state: $j_b(\theta) \leqq j_q$

$$\tau(\theta) = -\{c_a + \sigma(\theta)\tan\phi\}[\![1 - \exp[-a\{j_q - j_b(\theta)\}]]\!]$$

Thence, the axle load W, the braking torque Q_b and the apparent braking force T_{b0} acting on the tire which is descending a slope of angle β during braking action as shown in Figure 3.21

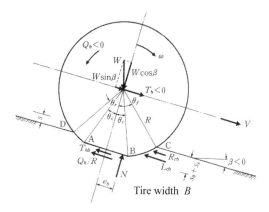

Figure 3.21. The various forces acting on a tire descending down sloped terrain during braking action.

can be computed by use of the following equations.

$$W \cos \beta = BR \left[\int_{\theta_C}^{\theta_f} \{\sigma(\theta) \cos \theta + \tau(\theta) \sin \theta\} \, d\theta + \int_{-\theta_C}^{\theta_C} p_g \cos\{\delta(\theta)\} \cos \theta \, d\theta \right.$$

$$\left. + \int_{-\theta_r}^{-\theta_C} \{\sigma(\theta) \cos \theta + \tau(\theta) \sin \theta\} \, d\theta \right] \tag{3.61}$$

$$Q_b = BR^2 \left[\int_{\theta_C}^{\theta_f} \tau(\theta) \, d\theta + \int_{-\theta_C}^{\theta_C} \tau(\theta) \cos \theta \, d\theta + \int_{-\theta_r}^{-\theta_C} \tau(\theta) \, d\theta \right] \tag{3.62}$$

$$T_{b0} = BR \left[\int_{\theta_C}^{\theta_f} \{\tau(\theta) \cos \theta - \sigma(\theta) \sin \theta\} d\theta + \int_{-\theta_C}^{\theta_C} \tau(\theta) \cos \theta \, d\theta \right. \tag{3.63}$$

$$\left. + \int_{-\theta_r}^{-\theta_C} \{\tau(\theta) \cos \theta - \sigma(\theta) \sin \theta\} \, d\theta \right]$$

In the above equation, the apparent effective braking force T_{b0} can be expressed as the difference between the drag T_{hb}, and the sum of the compaction resistance R_{cb} and the slope resistance $W \sin \beta$ as follows:

$$T_{b0} = T_{hb} - R_{cb} - W \sin \beta \tag{3.64}$$

where

$$T_{hb} = BR \int_{-\theta_r}^{\theta_f} \tau(\theta) \cos \theta \, d\theta \tag{3.65}$$

$$R_{cb} = BR \left[\int_{\theta_C}^{\theta_f} \sigma(\theta) \sin \theta \, d\theta + \int_{-\theta_r}^{-\theta_C} \sigma(\theta) \sin \theta \, d\theta \right] \tag{3.66}$$

In this case, the amount of sinkage s_s is not considered. The drag T_{hb} is given as the integral of $\tau(\theta)\cos \theta$ acting in the opposite direction to the moving direction of the tire and the compaction resistance R_{cb} e.g. the land locomotion resistance for the static amount of sinkage s_0 is given as the integral of $\sigma(\theta)\sin\theta$ acting in the same direction.

The braking force is given as Q_b/R. As shown in the previous Eq. (2.83), the horizontal component of the braking force $(Q_b/R)_h$ on a flat terrain equals the sum of the total compaction resistance L_b {e.g. the land locomotion resistance calculated considering the amount of slip sinkage} and the actual effective braking force T_b.

The amount of slip sinkage s_s at an arbitrary point X on the terrain can be calculated using Eq. (3.35) by substituting the amount of slippage j_s of the soil into Eq. (2.100). The depth of rut s that develops after the passage of the tire can be calculated by use of the previous Eq. (2.55) by substituting in the static amount of sinkage s_0, the amount of rebound u_0 given in the previous Eqs. (3.20) and (3.21) and the amount of slip sinkage s_s. Thence, the total compaction resistance L_{cb} can be determined, in the manner already mentioned in connection with the previous Eq. (2.103), as follows:

$$L_{cb} = k_1 \xi B \int_0^s z^{n_1} dz \tag{3.67}$$

The actual effective braking force T_b can be calculated as the difference between the tangential component of the braking force in the direction of the sloped terrain surface $(Q_b/R)_s$, and the sum of above mentioned total compaction resistance L_{cb} and the slope resistance $W\sin\beta$. This relation can be represented as:

$$T_b = \left(\frac{Q_b}{R}\right)_s - L_{cb} - W\sin\beta \tag{3.68}$$

When the amount of slip sinkage is not considered, the normal ground reaction N to the sloping terrain and the apparent amount of eccentricity e_{b0} may be given by:

$$N = W\cos\beta \tag{3.69}$$

$$e_{b0} = \frac{BR^2}{W\cos\beta} \left\{ \int_{\theta_c}^{\theta_f} p(\theta)\cos\theta\cos(\theta-\delta)\sin\theta\, d\theta \right.$$

$$\left. + \cos\theta_c \int_{-\theta_c}^{\theta_c} p(\theta)\cos\delta\, d\theta + \int_{-\theta_r}^{-\theta_c} p(\theta)\cos\theta\cos(\theta-\delta)\sin\theta\, d\theta \right\} \tag{3.70}$$

For full accuracy however, the actual amount of eccentricity e_b should be modified to allow for the effects of slip sinkage.

The effective input energy E_1 supplied from the braking torque Q_b acting on a tire can be equated to the sum of a number of output energy components: namely, a compaction energy component E_2 – associated with the development of a rut under the tire, a slippage energy component E_3 – associated with the development of shear deformation at the interface between tire and terrain, an effective braking force energy component E_4 e.g. the braking work and a potential energy component E_5 that comes into play on sloping terrains.

When the radius vector of the tire rotates from an entry angle θ_f at the beginning of the contact part of the tire-terrain interface through to a central angle $\theta_f + \theta_r$, the moving distance of the tire is equal to $R(\theta_f + \theta_r)/(1 + i_b)$ during a rotation of $R(\theta_f + \theta_r)$. The individual energy components, either generated or absorbed during the rotation of the tire, can then be calculated as follows:

$$E_1 = Q_b(\theta_f + \theta_r) \tag{3.71}$$

$$E_2 = \frac{L_{cb}R(\theta_f + \theta_r)}{1 + i_b} \tag{3.72}$$

$$E_3 = \frac{Q_b(\theta_f + \theta_r)i_b}{1 + i_b} \tag{3.73}$$

$$E_4 = \frac{T_bR(\theta_f + \theta_r)}{1 + i_b} \tag{3.74}$$

$$E_5 = \frac{WR(\theta_f + \theta_r)\sin\beta}{1 + i_b} \tag{3.75}$$

Following this, the production or expenditure or energy per unit time for each component can be calculated given knowledge of the peripheral speed for the wheel $R\omega$ and the moving speed V in the direction of the sloping terrain through use of the following equations:

$$E_1 = Q_b\omega = \left(\frac{Q_b}{R}\right)_s V(1 + i_b) \tag{3.76}$$

$$E_2 = \frac{L_{cb}R\omega}{1 + i_b} = L_{cb}V \tag{3.77}$$

$$E_3 = \frac{Q_b\omega i_b}{1 + i_b} = \left(\frac{Q_b}{R}\right)_s i_b V \tag{3.78}$$

$$E_4 = \frac{T_bR\omega}{1 + i_b} = T_bV \tag{3.79}$$

$$E_5 = \frac{WR\omega\sin\beta}{1 + i_b} = WV\sin\beta \tag{3.80}$$

Substituting the previous Eq. (3.68) into the above Eq. (3.79), the requisite energy equilibrium balance can be confirmed.

The 'optimum effective braking force' T_{bopt} during braking action can be defined as the effective braking force T_b at the 'optimum skid' i_{bopt} when the effective input energy $|E_1|$ takes a maximum value for a constant tire moving speed V. Thence, the braking efficiency E can be calculated using the previous Eq. (2.114).

3.4   KINEMATIC EQUATIONS OF A WHEEL

So far, the traffic performances of a rigid wheel and a flexible tired wheel during driving and braking action have been analysed assuming a constant peripheral speed $R\omega$ and a constant moving speed of tire V. These analyses were developed on the assumption that these wheels roll at a uniform speed and at a constant slip ratio or skid. In the more common, non-steady state condition however, the driving and braking torques $Q_d(t)$, $Q_b(t)$, and the effective driving and braking forces $T_d(t)$, $T_b(t)$ vary with time t. Similarly, the slip ratio i_d and the skid i_b will vary with time in correspondence with variations in the peripheral speed $R\omega$ and the moving speed V of the wheel.

Since the coefficients of driving and braking resistance μ_d, μ_b vary with time, these wheels will roll generally with either an accelerated speed or with a decelerated speed.

In this section, the fundamental kinematic equations of the straight forward motion and the rotational motion of the wheel will be analysed and developed for the pure rolling mode and for the driving and braking modes of operation of a wheel.

To begin these analyses, it is noted that the kinematic equations for the pure rolling state of the wheel can be developed in terms of the moment of inertia I and gravitational acceleration g as follows:

$$I\frac{d\omega}{dt} = -(\mu_r l_r + e_r)W \tag{3.81}$$

$$\frac{W}{g} \cdot \frac{dV}{dt} = \mu_r W - T_r \tag{3.82}$$

where $\mu_r(<0)$ is the coefficient of rolling friction, l_r is the normal distance between the application point of the ground reaction B_r acting in the direction of terrain surface and the central longitudinal axis as shown in the previous Figure 2.15. The factor e_r is the amount of eccentricity of the normal ground reaction N whilst T_r is the rolling resistance of the wheel.

When a wheel is drawn under conditions such that the rolling resistance T_r is always controlled to be equal to $\mu_r W$, the rate of change of velocity dV/dt becomes zero in the above equation and as a consequence the wheel rolls with a uniform speed under a constant skid $i_r(<0)$. Then, the angular velocity ω of the wheel becomes constant so that $d\omega/dt = 0$ and the condition $e_r = -\mu_r l_r (\mu_r < 0)$ is satisfied {as shown in the previous Eq. (2.80)}.

When a wheel is drawn by a T_r that is greater than $\mu_r W$, dV/dt becomes less than zero and consequently the wheel rolls with a decelerated speed. Since $B_r = \mu_r W$ increases with increments of skid, $d\omega/dt$ becomes less than zero in the above equation. Under these circumstances, both the moving speed V and the angular velocity ω decrease and the wheel decelerates at the same time.

On the other hand, when the wheel is drawn by a T_r that is smaller than $\mu_r W$, the factor dV/dt becomes larger than zero and as a consequence the wheel rolls with an accelerated speed. Since $B_r = \mu_r W$ decreases with decrements of skid, the parameter $d\omega/dt$ becomes larger than zero in the above equation so that both the moving speed V and the angular velocity ω increase and the wheel accelerates at the same time.

As a next step, the kinematic equation of the tire during driving action can be established for an applied driving torque as follows:

$$I\frac{d\omega}{dt} = Q_d(t) - (\mu_d l_d + e_d)W \tag{3.83}$$

$$\frac{W}{g} \cdot \frac{dV}{dt} = \mu_d W - T_d(t) \tag{3.84}$$

where μ_d is the coefficient of driving resistance, l_d is the normal distance between the application point of the ground reaction B_d acting in the direction of terrain surface and the central longitudinal axis as shown in the previous Figure 2.8(c). The factor e_d is the amount of eccentricity of the normal ground reaction N and $T_d(t)$ is the actual effective driving force. Additionally, μ_d is a factor that can be generally calculated from the previous Eqs. (2.25) and (3.37) and can be predicted for sloping terrains by the following expression:

$$\mu_d = \frac{B_d}{W} = \left(\frac{Q_d}{RW}\right)_s - \frac{L_{cd}}{W} - \sin\beta \tag{3.85}$$

When a wheel is drawn so that the effective driving force $T_d(t)$ is always controlled such that $B_d = \mu_d W$, the parameter dV/dt becomes zero in the above Eq. (3.84) and as a result the wheel rolls with a uniform speed at a constant slip ratio i_d. Thence, the driving torque $Q_d(t)$ becomes equal to $(\mu_d l_d + e_d)W$ so that $d\omega/dt = 0$, e.g. the angular velocity ω becomes a constant value. This results in the following equation;

$$\omega = \frac{V}{R(1 - i_d)} \tag{3.86}$$

where μ_d, l_d and e_d are given as the function of slip ratio i_d as mentioned previously. When $Q_d(t)$ increases or decreases {rather than having a value of $(\mu_d l_d + e_d)W$ for a constant moving speed V}, $d\omega/dt$ takes on a positive or negative value with the result that the circumferential speed of the wheel $R\omega$ either increases or decreases. Similarly, since the values of μ_d, l_d, e_d and $T_d(t)$ vary with increment or decrement of slip ratio i_d the force equilibrium of the system can be maintained. On the other hand, when $T_d(t)$ increases or decreases, rather than having a value of $\mu_d W$, for a constant peripheral rotational speed $R\omega$, dV/dt takes on negative or positive values so that the moving speed of the wheel decreases or increases. Under these circumstances, moment equilibrium is maintained through corresponding variation of μ_d, l_d, e_d and $Q_d(t)$. Likewise, when $Q_d(t)$ and $T_d(t)$ increase or decrease simultaneously, μ_d, l_d, e_d vary with V and $R\omega$ and force and moment balances are continuously maintained.

Supplementally, the kinematic equations of a tire during braking action can be set-up for a braking torque Q_b as follows:

$$I\frac{d\omega}{dt} = Q_b(t) - (\mu_b l_b + e_b)W \tag{3.87}$$

$$\frac{W}{g} \cdot \frac{dV}{dt} = \mu_b W - T_b(t) \tag{3.88}$$

where $\mu_b(<0)$ is the coefficient of braking resistance, l_b is the normal distance between the application point of the ground reaction B_b acting in the direction of the terrain surface and the central longitudinal axis as shown in the previous diagram, Figure 2.15(b). The parameter e_b is the amount of eccentricity of the normal ground reaction N and $T_b(t)$ is the actual effective braking force. Typically, the factor, μ_b can be calculated from the previous Eqs. (2.81) and (3.68). Predicted values can be obtained for sloping terrains using the following equation

$$\mu_b = \frac{B_b}{W} = \left(\frac{Q_b}{RW}\right)_s - \frac{L_{cb}}{W} - \sin\beta \tag{3.89}$$

When a tire is braked under conditions such that the effective braking force $T_b(t)$ is always controlled to have a magnitude $B_b = \mu_b W$, then dV/dt becomes zero in the above Eq. (3.88) and as a consequence one can see that the wheel rolls with an uniform speed with a constant amount of skid i_b. Under these circumstances, the braking torque $Q_b(t)$ becomes $(\mu_b l_b + e_b)W$, so that $d\omega/dt = 0$ e.g. the angular velocity ω takes on a constant value as described by the following equation:

$$R\omega = (1 + i_b)V \tag{3.90}$$

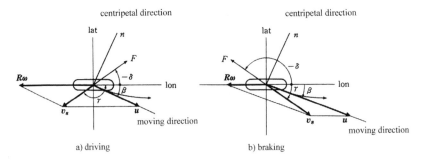

a) driving   b) braking

Figure 3.22. Friction force F acting on tire and slip velocity at cornering state.

where μ_b, l_b and e_b are functions of the amount of skid as mentioned previously. When $Q_b(t)$ increases or decreases rather than having a value equal to $(\mu_b l_b + e_b)W$ for a constant moving speed V, $d\omega/dt$ takes on a positive or negative value and as a consequence the circumferential speed of the wheel $R\omega$ increases or decreases. Because the values of μ_b, l_b, e_b and $T_b(t)$ vary with increment or decrement in the amount of skid i_b the system force equilibrium can be maintained. Alternatively, when $T_b(t)$ increases or decreases rather than having a value $\mu_b W$ for a constant peripheral rotational speed $R\omega$, dV/dt takes a negative or positive value such that the moving speed of the wheel decreases or increases. Accordingly, moment equilibrium can be maintained by variations in μ_b, l_b, e_b and $Q_b(t)$. Under conditions where $Q_b(t)$ and $T_b(t)$ increases or decreases simultaneously, μ_b, l_b, e_b vary with V and $R\omega$ such that the force and moment balances are always maintained.

The curves of $\mu_d - i_d$ and $\mu_b - i_b$ of a tire running on a soft terrain depend principally on the soil properties, the materials that comprise the tread rubber of the tire and the roughness of the tire surface. Curves of these parameters can be determined either from a simulation analysis that uses experimentally determined terrain-tire system constants or else can be directly determined from actual land locomotion tests on a tire during driving and braking action.

3.5   CORNERING CHARACTERISTICS

When a tire turns in some direction to the straight-line direction, the tire rolls but has associated with this process a lateral slippage at some angle β to the moving direction. This angle β is called the 'angle of lateral slippage β of the tire'. It is defined as the angle between the longitudinal direction and the moving direction of the tire as illustrated in Figure 3.22.

In the left-hand and right-hand diagrams (a) and (b), the directions of the friction force F acting on a tire during cornering action in driving and braking state are shown relative to the corresponding velocity vectors. For a velocity vector u in the moving direction of the tire and a velocity vector $R\omega$ in the rotational direction, a slippage velocity vector v_s can be expressed as $R\omega + u$. The angles between F, u, v_s and the longitudinal direction of the tire can be designated by the variables δ, β, γ respectively. The sense of these variables is defined such that clockwise angles represent positive values.

The longitudinal slip ratio or skid i_{lon} of a tire can be defined:
For the driving state where $0 \leqq u \cos \beta < R\omega$ as:

$$i_{lon} = 1 - \frac{u \cos \beta}{R\omega} \tag{3.91}$$

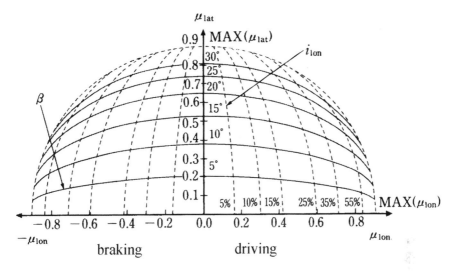

Figure 3.23. Relationship between longitudinal and lateral coefficients of friction μ_{lon} and μ_{lat} for various longitudinal slip ratio i_{lon} and angle of lateral slippage β [24].

For the braking state where $R\omega \leqq u\cos\beta$ as:

$$i_{lon} = \frac{R\omega}{u\cos\beta} - 1 \tag{3.92}$$

In the above equations, i_{lon} takes on a positive value for the driving state and negative one for the braking state.

In a similar vein, a lateral slip ratio i_{lat} for a tire can be defined as:

$$i_{lat} = -\frac{u\sin\beta}{u} = -\sin\beta \tag{3.93}$$

for positive values of the lateral slippage angle β. In this case, the lateral slip ratio i_{lat} takes a negative value as a consequence of the counterclockwise left turning of the tire. If the coefficient of friction μ is divided into a longitudinal component μ_{lon} and a lateral component μ_{lat}, the relationship between these components can be expressed as a group of elliptical curves. Figure 3.23 [23, 24] shows a group of such curves developed around the two parameters of longitudinal slip ratio i_{lon} and angle of lateral slippage β. It can be seen here that the lateral coefficient of friction μ_{lat} takes a maximum value of $(\mu_{lat})_{max}$ at $\mu_{lon} = 0$ for each value of β. Further, this value can be expressed as an exponential function of β as in the following equation:

$$(\mu_{lat})_{max} = MAX(\mu_{lat})\{1 - \exp(-k_1\beta)\} \tag{3.94}$$

where $MAX(\mu_{lat})$ is the maximum value of μ_{lat} for the whole range of the values of β and i_{lon}.

Similarly, the longitudinal coefficient of friction μ_{lon} takes a maximum value of $(\mu_{lon})_{max}$ at $\mu_{lat} = 0$ for each value of i_{lon} and can be expressed as an exponential function of i_{lon} as follows:

$$(\mu_{lon})_{max} = MAX(\mu_{lon})\{1 - \exp(-k_2 i_{lon})\} \tag{3.95}$$

where $MAX(\mu_{lon})$ is the maximum value of μ_{lon} for the whole range of the values of β and i_{lon}.

In the above two equations, the coefficients k_1 and k_2 vary with the soil properties and the tread pattern of the tire.

Further, the group of elliptical curves that are shown by the solid lines in the figure, can be expressed as follows:

$$\left\{\frac{\mu_{lat}}{(\mu_{lat})_{max}}\right\}^2 + \left\{\frac{\mu_{lon}}{MAX(\mu_{lon})}\right\}^2 = 1 \qquad (3.96)$$

and another group of elliptical curves, shown by the dotted lines, can be expressed as follows

$$\left\{\frac{\mu_{lat}}{MAX(\mu_{lat})}\right\}^2 + \left\{\frac{\mu_{lon}}{(\mu_{lon})_{max}}\right\}^2 = 1 \qquad (3.97)$$

An approximately linear relationship can be established between the coefficient of friction μ and the slip ratio i of a tire running on a hard terrain.
Namely:

$$\mu_{lon} = c_{lon}i_{lon} \qquad (3.98)$$

$$\mu_{lat} = c_{lat}i_{lat} \qquad (3.99)$$

Likewise, the resultant force F applied to a tire can be divided into a longitudinal component F_{lon} and a lateral component F_{lat} as follows:

$$F_{lon} = \mu_{lon}W \qquad (3.100)$$

$$F_{lat} = \mu_{lat}W \qquad (3.101)$$

Further, the resultant force F can be sub-divided into a drag force F_B acting in the moving direction of tire and a cornering force F_C acting in the centrifugal direction as shown in Figure 3.24:

$$F_B = F_{lon}\cos\beta - F_{lat}\sin\beta = (c_{lon}i_{lon}\cos\beta - c_{lat}i_{lat}\sin\beta)W \qquad (3.102)$$

$$F_C = F_{lon}\sin\beta + F_{lat}\cos\beta = (ci_{lon}\sin\beta + c_{lat}i_{lat}\cos\beta)W \qquad (3.103)$$

In this diagram, the symbol M is the self-aligning torque that occurs as a result of the point of application of the cornering force F_C moving from the center of the tire towards the rear of the tire.

Figure 3.25 shows the shape of the contact part of the tire during cornering action in the driving state. Initially, the tread of the tire touches the terrain at the beginning contact point l and moves along the line l–m during the rolling of the tire.

In this situation, the cornering force F_C increases with the lateral deformation of the tire until it reaches a maximum frictional resistance when slip occurs reversely due to an elastic deformation of the rubber tread. Then, the tire tread kicks away from the point m and recovers suddenly to the end contact point n. The interval l–m is referred to as the cohesive

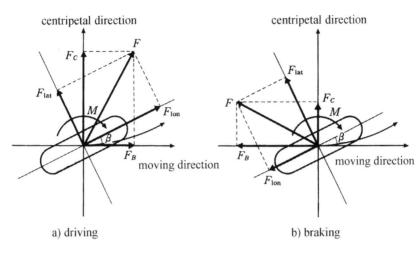

a) driving   b) braking

Figure 3.24. Forces acting on tire.

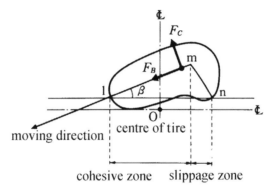

Figure 3.25. Shape of contact area cornering state (during driving action).

zone and that of m–n as the slippage zone. In this case, the drag force F_B can be calculated from the driving or braking torque and the land locomotion resistance, and the cornering force F_C can be computed as the sum of the static frictional force and the dynamic frictional force acting on the contact part of the tire. Sakai [25] has developed a detailed mechanical model of the process of transitioning from the static frictional zone to the dynamic one.

As mentioned previously, the application point of F_C deviates to the rear part of the tire and away from the central line of the tire which is normal to the terrain surface. From this a self-aligning torque M develops. Gough [26] measured the relationship between F_C and M for a 165-SR-13 tire with air pressure of 167 kPa for various values of axle load W. The results he obtained are shown in Figure 3.26.

The value of M takes a maximum value for the range of $\beta = \pi/45 \sim \pi/30$ rad and after that it decreases gradually. The value of M increases linearly with increments in F_C for a range of small values of β but tends to decrease for a range of large values of β. The Newmatic trail NT is the amount of eccentricity that exists between the application point

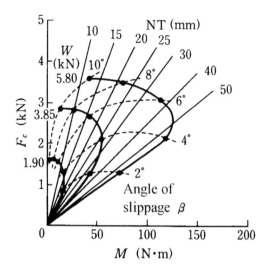

Figure 3.26. Relationship between cornering force F_c and self aligning torque M [26].

of F_C and the central line of the tire normal to the terrain surface. This parameter can be calculated as follows:

$$NT = \frac{M}{F_C} \tag{3.104}$$

The ratio of the cornering force F_C to the angle of slippage β for the range of small values of β, i.e. F_C/β is designated as the 'Cornering power'. The ratio of the cornering power to the axle load of the tire, i.e. $F_C/W\beta$ is designated the 'Coefficient of cornering'. Both these terms may be used to describe the cornering characteristics of a tire.

3.6   DISTRIBUTION OF CONTACT PRESSURE

In this section, some illustrative results relative to measured contact pressure distributions at the tire-terrain interface are presented. Krick [13] measured the distribution of the contact pressure that existed on the surface of a pneumatic tire that was rolling on a loose accumulated sandy loam terrain, of water content 19%. He obtained his measurements by use of a three dimensional transducer. Figure 3.27 shows the distributions of contact pressure e.g. the normal stress σ and the shear resistance τ measured for a tire during driving state operations. The axle load was 5.34 kN and the slip ratio 10%.

In this case, the shear resistance distribution shows a triangular shape which increases linearly with distance from the beginning point of the contact of the tire with the terrain. On the other hand, the normal stress follows a parabolic shape. In the diagram, the curve denoted by the symbol 1 refers to the results measured at the center of the tire width whilst curve 4 refers to the results measured at the edge part of the tire. It is noted that the normal stress shows some stress concentration at the edge part of the tire.

Figure 3.28 shows the distributions of contact pressure of a tire of axle load 5.34 kN operating in a driving state with a slip ratio of 40%. Under such conditions of large slip ratio values, it is observed that the distributions of normal stress and shear resistance show

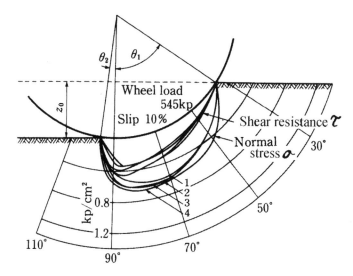

Figure 3.27. Contact pressure distributions under tire (slip ratio 10% during driving action) [13].

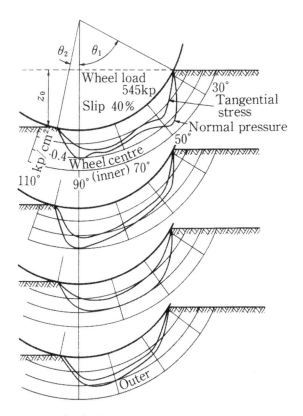

Figure 3.28. Contact pressure distributions acting on tire (slip ratio 40% during driving action) [13].

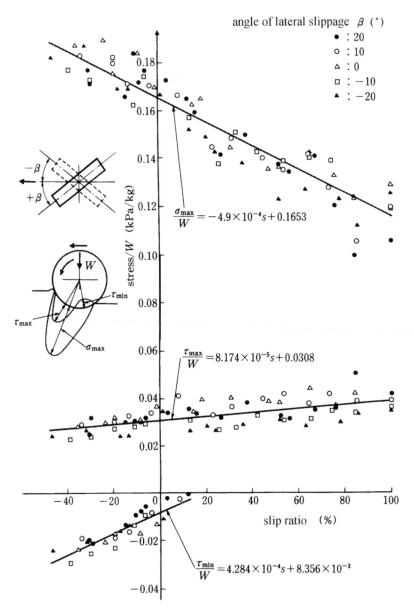

Figure 3.29. Relationship between ratio of maximum normal stress σ_{max}, maximum and minimum shear resistance τ_{max}, τ_{min} to axle load W and slip ratio s.

a somewhat flattened shape at the center but they also show a triangular shape in the entry and exit contact regions of the tire.

In somewhat similar studies, Oida et al. [27] measured values for the distributions of normal stress σ, longitudinal and lateral shear resistance τ acting on the contact part of a tire running on a standard sand. They did their work systematically for various combinations of longitudinal slip ratio s and angle of lateral slippage β. The sand had an angle of internal friction of 38 degrees. The tire was of type 4.50-5-4 PR and the air pressure was 49 kPa.

Figure 3.29 presents the results of this work in terms of three normalised parameters i.e. the maximum normal stress divided by the axle load σ_{max}/W, the maximum longitudinal shear resistance divided by the axle load τ_{max}/W, the minimum longitudinal shear resistance divided by the axle load τ_{min}/W. These normalised values are plotted against the longitudinal slip ratio s for various values of angle of lateral slippage β.

Some experimental relations that best-fit this data are as follows.

$$\frac{\sigma_{max}}{W} = -4.9 \times 10^{-4}s + 0.1653 \tag{3.105}$$

$$\frac{\tau_{max}}{W} = 8.174 \times 10^{-5}s + 0.0308 \tag{3.106}$$

$$\frac{\tau_{min}}{W} = 4.284 \times 10^{-4}s + 8.356 \times 10^{-3} \tag{3.107}$$

From these results, it is observed that the angle of lateral slippage β does not significantly affect the distributions of normal stress σ and shear resistance τ. The data also shows that the ratio of the angle that develops a maximum normal stress σ_{max} to the angle developing a maximum shear resistance τ_{max} for a specific entry angle can be expressed as a polynomial function of the longitudinal slip ratio s.

3.7   SUMMARY

In this chapter, the work of Chapter 2 on rigid drums has been extended to cover elastic and deformable drums and short cylinders operating over hard and deformable terrains. Because these systems are hard to handle mathematically and because three dimensional effects cannot reasonably be ignored for most rubber tired vehicles, this chapter has put together a mix of empirical and analytical functions that can be used in tired-vehicle behaviour prediction situations.

In this chapter, basically the same set of drum/terrain modelling situations that were covered in Chapter 2 with rigid wheels have been revisited with flexible wheels.

REFERENCES

1. Kommandi, G. (1975). Determination of the Peripheral force for Pneumatic Tyres Rolling on Concrete Surfaces. *Proc. 5th Int. Conf. ISTVS, Michigan, U.S.A., Vol. 2.* pp. 567–588.
2. Muro, T. & Enoki, M. (1983). Characteristics of Friction of OR Tyre at Cornering Site. *Memoirs of the Faculty of Engineering, Ehime University, Vol. X, No.2,* pp. 285–297, (In Japanese).
3. Yong, R.N., Boonsinsuk, P. & Fattah, E.A. (1980). Tyre Flexibility and Mobility on Soft Soils. *J. of Terramechanics, 17, 1,* 43–58.
4. Fujimoto, Y. (1977). Performance of Elastic wheels on Yielding Cohesive Soils. *J. of Terramechanics, 14, 4,* 191–210.
5. Forde, M.C. (1978). An Investigation into Rubber Wheel Mobility on London Clay and Cheshire Clay. *Proc. 6th Int. Conf. ISTVS, Vienna, Austria, Vol. 2.* pp. 587–642.
6. Yong, R.N., Boonsinsuk, P. & Fattah, E.A. (1980). Tyre Load Capacity and Energy Loss with Respect to Varying Soil Support Stiffness. *J. of Terramechanics, 17, 3,* 131–147.
7. Yong, R.N., Fattah, E.A. & Boonsinsuk, P. (1978). Analysis and Prediction of Tyre–Soil Interaction and Performance using Finite Elements. *J. of Terramechanics, 15, 1,* 43–63.

8. Abeels, P. (1981). Studies of Agricultural and Forestry Tyres on Testing Stand, Basis for Data Standardizations. *Proc. 7th Int. Conf. ISTVS Calgary, Canada, Vol. 2*, pp. 439–453.

9. Fujimoto, Y. (1977). Performance of Elastic Wheels on Yielding Cohesive Soil. *J. of Terramechanics, 14, 4*, 191–210.

10. Harlow, S., Krutz, G., Liljedahl, J.B. & Parsons, S. (1981). Relation between Tractor Tyre Wear and Lug Forces. *Proc. 7th Int. Conf. ISTVS, Calgary, Canada. Vol. 2.* pp. 645–662.

11. Muro, T. (1983). Characteristics of Wear life of Heavy Dump Truck Tyre. *Proc. of JSCE*, No. 336, pp. 149–157, (In Japanese).

12. Turnage, G.W. (1978). A Synopsis of Tire Design and Operational Considerations Aimed at Increasing In-Soil tire Drawbar Performance. *Proc. 6th Int. Conf. ISTVS Vienna, Austria, Vol. 2*, pp. 759–810.

13. Krick, G. (1969). Radial and Shear Stress Distribution under Rigid Wheels and Pneumatic Tires Operating on Yielding Soils with Consideration of tire Deformation. *J. of Terramechanics, 6, 3*, 73–98.

14. Oida, A. & Triratanasirichai, K. (1988). Measurement and Analysis of Normal, Longitudinal and Lateral Stresses in Wheel-Soil Contact Area. *Proc. 2nd Asia-Pacific Conf. ISTVS, Bangkok, Thailand.* pp. 233–243.

15. Söhne, W. (1953). Druckverteilung im Boden und Bodenverformung unter Schleppereifen. *Glundlagen der Landtechnik, Heft 5*, pp. 49–63.

16. Muro, T. (1985). Compaction Phenomena due to Vehicle Transportation. *Tsuchi-to-Kiso, Vol. 33, No. 9*, pp. 33–38, (In Japanese).

17. Bolling, I. (1981). Verticalspannungsverteilungen im Boden unter Luftreifen, *Proc. 7th Int. Conf. ISTVS Calgary, Canada. Vol. 2*, pp. 497–529.

18. Bekker, M.G. Prediction of Design and Performance Parameters in Agro-Forestry Vehicles. *National Research Council of Canada*, Report No. 22880.

19. Schwanghart, H. (1991). Measurement of Contact Area, Contact Pressure and Compaction under Tires in Soft Soil. *J. of Terramechanics, 28, 4*, 309–318.

20. Karafiath, L.L. & Nowartzki, E.A. (1978). *Soil Mechanics for Off-Road Vehicle Engineering.* pp. 355–427, Transtech Publications.

21. Blaszkiewicz, Z. (1990). A Method for the Determination of the Contact area between a Tyre and the Ground. *J. of Terramechanics, 27, 4*, 263–282.

22. Wong, J.Y. (1989). *Terramechanics and Off-Road Vehicles.* pp. 214–241. Elsevier.

23. Grecenko, A. (1975). Some Applications of the Slip and Drift Theory of the Wheel. *Proc. 5th Int. Conf. ISTVS, Michigan, U.S.A.* Vol. 2, pp. 449–472.

24. Crolla, D.A. & El-Razaz, A.S.A. (1987). A Review of the Combined Lateral and Longitudinal Force Generation of Tyres on Deformable Surfaces. *J. of Terramechanics, 24, 3*, 199–225.

25. Sakai, H. (1969). Theoretical Considerations of the Effect of Braking and Driving Force on Cornering Force. *Automotive Engineering, 23, 10*, 982–988, (In Japanese).

26. Gough, V.E. (1954). Cornering Characteristics of Tyres. *Aut. Engr., Vol. 44, No. 4.*

27. Oida, A., Satoh, A., Itoh H. & Triratanasirichai, K. (1991). Three-Dimensional Stress Distributions on a Tire-Sand Contact Surface. *J. of Terramechanics, 28, 4*, 319–330.

EXERCISES

(1) Suppose that an off-road-tire of radius of $R = 1.2$ m is standing on a hard terrain and that it is sustaining an axle load of $W = 100$ kN. Under these conditions the amount of deformation of the tire f is observed to be 3 cm. The radius of the tread crown r of the tire is 30 cm. Assuming that the tire deforms only at the contact tire/terrain interface and that the contact pressure equals the tire air pressure, calculate the contact area A and the lengths of the major axis $2a$ and the minor axis $2b$ of the ellipse.

(2) Suppose that an off-the-road tire of radius $R = 1.20$ m is running on a hard terrain, in a pure rolling state, and that it sustains an axle load of $W = 100$ kN. If the effective radius

of rotation R_{e0} is 1.16 m and the coefficient of rolling friction μ_r is 0.20, calculate the effective braking force T_b and the amount of eccentricity e_r of the ground reaction.

(3) Imagine that an off-the-road tire of radius of $R = 1.30$ m is running during driving action on a hard terrain whilst it is sustaining an axle load of $W = 100$ kN. If the applied driving torque Q_d is 100 kNm, the effective radius of rotation R_{e0} is 1.25 m, and the coefficient of rolling friction μ_d is 0.20, calculate the coefficient of driving resistance μ_d, the effective driving force T_d, and the amount of eccentricity e_d of the ground reaction.

(4) Imagine that an off-the-road tire of radius of $R = 1.20$ m, width of $B = 0.5$ m is running during driving action on a weak sandy terrain and whilst it is doing this it is sustaining an axle load of $W = 20$ kN. If the height of the tire H is 0.6 m, the gradient of cone index G of the terrain is 12 N/cm^3, and the amount of deformation δ of the tire is 3.0 cm, calculate the effective tractive effort T_{20} and the tractive efficiency η_{20} at a slip ratio of $i_d = 20\%$ from the wheel mobility number N_s.

(5) Consider an off-the-road tire of radius of $R = 1.0$ m which is running during braking action on a hard terrain while it is sustaining an axle load of $W = 100$ kN. Given an applied braking torque Q_b of 80 kNm, an effective radius of rotation R_{eb} is 0.96 m, and a coefficient of rolling friction $\mu_r = 0.15$, calculate the coefficient of braking resistance μ_b, the effective braking force T_b and amount of eccentricity e_b of the ground reaction.

(6) An off-the-road tire of radius of $R = 1.2$ m and width of $B = 0.8$ m is running during driving action on a soft clayey terrain while sustaining an axle load of $W = 10$ kN. Suppose that the height of the tire H is 0.6 m, the cone index C of the terrain is 15.6 N/cm^2, and the amount of deformation δ of the tire is 5 cm. Calculate the effective tractive effort T_{20} and the tractive efficiency η_{20} at a slip ratio of $i_d = 20\%$ from the wheel mobility number N_c.

(7) A 13.6/12-28 tire of diameter $D = 1310$ mm and width $B = 345$ mm is standing on a loose accumulated sandy loam terrain of water content $w = 15\%$. It is sustaining an axle load of $W = 17.9$ kN. The tire air pressure p is 150 kPa. Does the tire behave as a rigid wheel or as a flexibly tired wheel?

(8) A tire of radius of $R = 30$ cm, width of $B = 10$ cm sustaining the axle load $W = 3000$ N is descending a slope of $\beta = \pi/36$ rad in a pure rolling mode. The depth of rut produced is 5 cm and values of terrain-tire system constants equal to $k_1 = 4.5$ N/cm$^{n_1+1}$, $n_1 = 0.809$ and $\xi = 1$ have been obtained from a plate loading test. Calculate the effective braking force T_b in this case.

(9) An off-the-road tire is climbing up a slope of angle β during driving action. Demonstrate that the effective input energy per second E_1 is equal to the sum of the output energies per second, i.e. equal to the summation of the sinkage deformation energy E_2, the slippage energy E_3, the effective drawbar pull energy E_4, and the potential energy E_5.

(10) Suppose that a tire, which is sustaining an axle load of $W = 2.5$ kN, is cornering during driving action on a sandy terrain at a slip angle of $\beta = \pi/18$ rad and that a longitudinal slip ratio of $i_{lon} = 15\%$ prevails. Calculate the cornering force F_C and the drag force F_B in this case.

Chapter 4

Terrain-Track System Constants

Track-laying vehicles, as they manifest in modern machines such as bulldozers, army tanks, tracked mobile-cranes or snowmobiles comprise a major class of land-locomotion device. These vehicles interact with the ground though a track belt structure that can be thought of as existing in two quite distinctive forms – a rigid track form and a flexible track form. Because the interaction between a track belt and a terrain is fundamentally different in the rigid and the flexible belt types, the overall structure of a machine's track belt system will dictate the traffic performance of the machine on construction sites or on other types of on-road or off-road application.

A multiroller-type track belt [1] is an example of the essentially fully-rigid track belt type. In this arrangement, the movement of the track bush that mutually connects several track links is fully constrained by a structural girder. As a result of this constraint, no upper or lower movement of the track plates that are connected to the track links can occur. The distribution of contact pressure under a rigid track belt system is uniform and as a consequence rigid tracked vehicles can typically develop large drawbar pulls – especially on soft terrains.

On the other hand, most common bulldozers and tractors are examples of an essentially flexible track belt undercarriage system. In this arrangement, upper and lower movements of the track plates between several road rollers running on the track link can occur. As well, right and left hand lateral movements of the road rollers can take place. The distribution of contact pressure under such flexible track belt systems follows a wavy distribution with stress concentrations existing just under the road rollers. As a consequence of this, flexible track vehicles can be used for operations on rough terrain.

In the study of tracked machinery, a variety of parameters that collectively will be referred to as *terrain-track system constants* can be used to describe and predict the interactions that occur between a track belt structure and a terrain. The terrain-track constants have been developed through considerations of basic soil mechanics.

The terrain-track system constants are constants that can model the relations between contact pressure and amount of sinkage through use of a model-track-plate loading and unloading test. As well, the relations that exist between shear resistance, contact pressure and amount of slippage as well as the relations that exist between the amount of slip sinkage, contact pressure and the amount of slippage of the soil can be derived from a model-track-plate traction test. These constants are dependent on the structure of the track belt, the ratio of the grouser pitch to height and the properties of the terrain materials.

In what follows, the terrain-track system constants for silty loam, decomposed weathered granite soil and snow covered terrains will be investigated. As well, the size effect of the model-track on the constants will be studied.

4.1   TRACK PLATE LOADING TEST

The relations between the contact pressure p acting on a track belt and the static amount of sinkage s_0 vary with the terrain properties, the structure of the track belt and the size of the model-track. Because of these multiple factors it is necessary to carry-out loading tests on model-tracks equipped with variously sized track plates of given grouser height H and grouser pitch G_p for particular terrains.

The experimental measurements that cover this type of test generally take the following form.

For $p \leq p_0, s_0 \leq H$

$$p = k_1 s_0^{n_1} \tag{4.1}$$

For $p > p_0, s_0 > H$

$$p = p_0 + k_2 (s_0 - H)^{n_2} \tag{4.2}$$

where $p_0 = k_1 H^{n_1}$.

The two equations cover the situations where (a) the grouser penetrates the surface but the track-plane has not landed and (b) the grouser has penetrated and contact with the bottom plane of the model-track has occurred. As well, it is necessary to execute an unloading test on the model-track element from a start-point of an arbitrary amount of sinkage s_p. This test is required so as to determine the relations between the contact pressure p and the static amount of sinkage s_0 in the unloading phase since a process of unloading takes place after the pass of a road roller on a flexible track belt. For the unloading case, the form of the experimental equations is as follows:

For $p \leq p_0, s_0 \leq H$

$$p = k_1 s_p^{n_1} - k_3 (s_p - s_0)^{n_3} \tag{4.3}$$

For $p > p_0, s_0 > H$

$$p = p_0 + k_2 (s_p - H)^{n_2} - k_4 (s_p - s_0)^{n_4} \tag{4.4}$$

where the coefficients of sinkage k_1, k_2, k_3 and k_4, and the indices of sinkage n_1, n_2, n_3 and n_4 are the set of terrain-track system constants that are determined from the model-track-plate loading test.

4.2   TRACK PLATE TRACTION TEST

From a traction test on a model-track-plate, the relationship that exists between the shear resistance acting at a track-terrain interface, the contact pressure p and the amount of slippage j can be experimentally determined. Also, the relationship between the amount of slip sinkage s_s, the contact pressure p and the amount of slippage j_s of a soil can be determined.

The shear resistance τ of a soil that develops under a track belt can not be expressed simply by a cohesion and angle of internal friction expression that satisfies the Mohr-Coulomb's

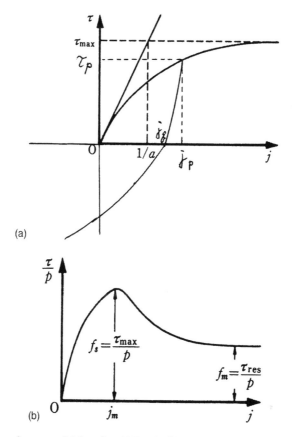

Figure 4.1. (a) Type of exponential function (A Type). (b) Hump type relationship (B Type) between shear resistance τ and amount of slippage j.

failure criterion [2]. The reason for this is that also involved is the failure pattern of the soil that is induced by the structure of track belt and the degree of mobilization of the shear strength of the soil that occurs due to the operation of various slippage processes.

In general, the shear resistance τ under a track belt can be expressed as a function of the apparent cohesion m_c, the apparent angle of shear resistance m_f, the contact pressure p and the amount of slippage j. As shown in Figure 4.1, the function can be classed into two representative types that reflect particular terrain properties. Thus, one can have an exponential type function (Type A) that passes through the origin and is asymptotic to a maximum value τ_{max}. The other type of expression is a Hump type function (Type B) which exhibits a distinctive peak value.

For loose accumulated sandy soils, remolded soft clayey soils and for normal consolidated clays, Janosi-Hanamoto [3] and others have proposed the following Type A function:

For $0 \leq j \leq j_p$ (traction state):

$$\tau = (m_c + m_f p)\{1 - \exp(-aj)\}$$

For $j_q < j < j_p$ (untraction state):

$$\tau = \tau_p - k_0(j_p - j)^{n0} \tag{4.5}$$

For $j \leq j_q$ (reciprocal traction state):

$$\tau = -(m_c + m_f p)[1 - \exp\{-a(j_q - j)\}]$$

where j_p is an arbitrary value for the amount of slippage j at the beginning of the untraction state, τ_p is the corresponding shear resistance at $j = j_p$, and j_q is the value of j when the shear resistance τ becomes zero in the untraction process. Further, the constant a is the coefficient of deformation divided by τ_{max}.

For a firmly compacted sandy soil and a hard terrain, the following Type B function has been proposed by Oida [4]:

$$\tau = f_m \cdot p \left\| 1 - \frac{\sqrt{1 - f_m/f_n} \exp\left[(j/j_m) \log\left\{ 1 + f_s/f_m \cdot \left(\sqrt{1 - f_s/f_m} - 1 \right) \right\} \right]}{\sqrt{1 - f_m/f_s} \, (1 - 2f_s/f_m) + 2f_s/f_m - 2} \right\|$$

$$\times \left\| 1 - \exp\left[(j/j_m) \log\left\{ 1 + f_s/f_m \cdot \left(\sqrt{1 - f_m/f_s} - 1 \right) \right\} \right] \right\| \tag{4.6}$$

where f_m is the ratio of the residual shear resistance and the contact pressure τ_{res}/p, f_s is the ratio of the maximum shear resistance and the contact pressure τ_{max}/p, and j_m is the amount of slippage corresponding to the maximum shear resistance τ_{max}. The above equation can also be applied for non-adhesive asphalt and concrete pavement.

For an overconsolidated clay, the following B type function was proposed by Bekker [5]:

$$\tau = (m_c + m_f p)$$

$$\times \frac{\exp\left\{ \left(-K_2 + \sqrt{K_2^2 - 1} \right) K_1 j \right\} - \exp\left\{ \left(-K_2 - \sqrt{K_2^2 - 1} \right) K_1 j \right\}}{\exp\left\{ \left(-K_2 + \sqrt{K_2^2 - 1} \right) K_1 j_m \right\} - \exp\left\{ \left(-K_2 - \sqrt{K_2^2 - 1} \right) K_1 j_m \right\}} \tag{4.7}$$

$$K_1 = \frac{1}{j_m} \cdot \frac{\log\left(-K_2 - \sqrt{K_2^2 - 1} \right)}{\sqrt{K_2^2 - 1}}$$

where K_1, K_2 are the coefficients of deformation of the soil and j_m is the amount of slippage corresponding to the maximum shear resistance.

The above mentioned coefficient e.g. m_c, m_f and a in Eq. (4.5), f_m, f_s and j_m in Eq. (4.6), and K_1, K_2 and j_m in Eq. (4.7) are the terrain-track system constants determined from the track-plate traction test.

Some further relations between the amount of slip sinkage s_s and the amount of slippage of soil j_s may also be deduced from the track plate traction test. The slip sinkage of the track belt occurs due to a dilatation phenomenon [6] that appears in concert with the shear deformation of the soil under a track belt. It is also due to the scratching effects of the grousers. In general, the amount of slip sinkage s_s measured at the rear end of the model-track-plate is given as a function of contact pressure p and amount of slippage j_s as follows:

$$s_s = c_0 p^{c_1} j_s^{c_2} \tag{4.8}$$

where the constant c_0 as well as the indices c_1, c_2 vary with the terrain properties and the structure of the track belt. These terrain-track system constants need to be determined from a series of track-plate traction tests.

4.3   SOME EXPERIMENTAL RESULTS

4.3.1   Effects of variation in grouser pitch–height ratio

The terrain-track system constants vary very considerably with the physical form of a particular track belt. Specifically, the ratio of the grouser pitch G_p to the grouser height H seems to be a very important parameter in determining the magnitude of the overall tractive action that develops under a track belt.

From traction tests [7] carried out on a model-track-plate which was equipped with standard T shaped rubber grousers and which was operating on a sandy terrain, it was observed that the traction force took on a maximum value when the G_p/H ratio was in the range $3 \sim 4$.

Figure 4.2 shows the tractive effort development mechanism for a model-track-plate which is equipped with standard T shaped grousers. With increasing amounts in slippage j of the model-track-plate, slip lines grow in the soil between the side planes of the grousers. As a consequence, tractive effort develops due to progressive failure along the slip lines which in turn develops shear resistance.

When the ratio of G_p/H takes on values less than around 2, the profile of the slip line in the soil between the grousers approaches the form of a straight line connecting each of the tips of the grousers. Under these conditions, the maximum tractive effort T of a model-track-plate having n individual grouser elements can be determined for a track width B, a contact pressure p, a coefficient of earth pressure at rest K_0, and an angle of soil shear resistance $\tan \phi$ and represented by the following equation:

$$T = nG_p Bp \tan \phi \left[1 + \frac{2HK_0}{B} \right] \tag{4.9}$$

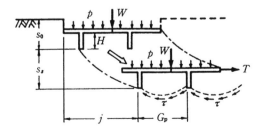

Figure 4.2. Mechanism of occurrence of slip lines and tractive effort.

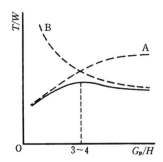

Figure 4.3. Relation between coefficient of traction T/W and grouser pitch-height G_p/H.

In actuality, the value of T can be presumed to decrease gradually with decrements in the grouser pitch G_p, because the base area of the grouser itself becomes relatively large with decreasing G_p and also because the frictional resistance between the grouser tip and the soil becomes less than the internal shear resistance of the soil.

When the ratio of G_p/H takes on values larger than around 4, the slip line in the soil between the grousers becomes a passive failure line which consists of a logarithmic spiral slip line and a straight slip line. This failure line develops in front of each grouser. Under these conditions, the maximum tractive effort T of a model-track-plate having n individual grouser elements can be approximated (for a coefficient of passive earth pressure K_p) by the following expression:

$$T = nHBK_pp + nG_pHpK_0 \tan \phi \tag{4.10}$$

In this case, the vertical load W acting on the model-track-plate is equal to the product nG_pBp. Also, by dividing the above equation by W, the following expression is obtained:

$$\frac{T}{W} = \frac{HK_k}{G_p} + \frac{HK_0 \tan \phi}{B} \tag{4.11}$$

where T/W decreases with increases in G_p.

Figure 4.3 shows the relations between the ratios T/W and G_p/H. In the Figure the dotted line A represents a modified Eq. (4.9). The dotted line B represents Eq. (4.11).

Under real world conditions, however, a somewhat lesser value of T/W than that suggested by the A or B lines occurs. The ratio T/W takes a maximum value at some value of G_p/H as shown by the solid line. It is noted however that this optimum grouser pitch–height ratio $(G_p/H)_{opt}$ varies with the particular properties of the terrain.

Likewise to the above, from traction tests [8] for a model-track-plate equipped with a set of standard T shaped steel grousers and operating on a remolded silty loam terrain, it was observed that the shear resistance of the soil τ_{70} at an amount of slippage $j = 70$ cm took on a maximum value at $G_p/H \cong 3.2$. Figure 4.4 shows the relationships between τ_{70} and G_p/H for three magnitudes of contact pressure $p = 7.1$, 10.1, and 12.3 kPa. The model-track-plate of width $B = 20$ cm was equipped with 7 individual T shaped steel grousers of height $H = 3.2$ cm. The water content and the cone index of the silty loam terrain were measured to be approximately 30% and 31 kPa respectively.

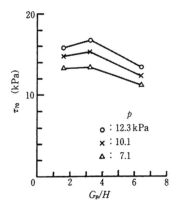

Figure 4.4. Relationship between shear resistance τ_{70} at amount of slippage 70 cm and grouser pitch–height ratio G_p/H.

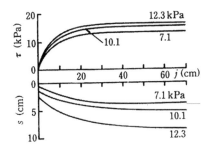

Figure 4.5. Relationship between shear resistance τ, amount of sinkage s and amount of slippage $j(G_p = 10.2 \text{ cm})$.

Table 4.1. Terrain-track system constants (silty loam terrain).

G_p (cm)	5.1	10.2	20.4
k_2 (N/cm$^{n_2+2}$)	10.50	10.17	9.85
n_2	0.395	0.395	0.395
m_c (kPa)	10.00	9.020	8.330
m_f	0.471	0.620	0.401
a (1/cm)	0.140	0.150	0.130
c_0 (cm$^{2c_1-c_2+1}$/N^{c_1})	0.139	0.100	0.105
c_1	0.860	0.855	0.754
c_2	0.425	0.466	0.403

Figure 4.5 plots the relations between the shear resistance τ, the amount of sinkage $s = s_0 + s_s$ and the amount of slippage j for a grouser pitch of $G_p = 10.2$ cm. Table 4.1 gives a summary of all the terrain-track system constants, including the model-track-plate loading test results, for three values of grouser pitch $G_p = 5.1$, 10.2 and 20.4 cm.

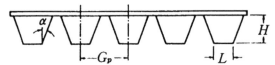

Width $B = 25$ cm, grouser pitch $Gp = 14.6$ cm,
grouser height $H = 6.5$ cm, base length $L = 2, 3, 4.5$ cm,
trim angle $\alpha = \pi/6$ rad

Figure 4.6. Shape and dimension of track model plate.

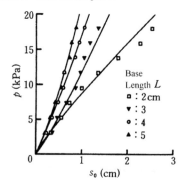

Figure 4.7. Relationship between contact pressure p and static amount of sinkage s_0.

4.3.2   Results for a decomposed granite sandy terrain

In this section some results for both a track plate loading and a traction test are presented for a model-track-plate equipped with equilateral trapezoidal rubber grousers that is acting on a loose accumulated decomposed weathered granite sandy terrain.

The test decomposed granite sandy soil has a specific gravity of 2.66, an average grain size of 0.78 mm, a coefficient of uniformity of 2.0, a maximum density of 1.68 g/cm^3, and a minimum density of 1.31 g/cm^3. The air dried decomposed granite sandy soil was deposited by free-falling the material from a height of 1.0 m and filling a bin uniformly until a depth of 30 cm was attained. The test was undertaken in a large soil bin of length 270 cm, width 90 cm and height 60 cm.

The relative density of the test terrain prepared in this way was 44.0%, the water content was 2.38%, the dry density was 1.44 g/cm^3 and the cone index was 95.1 kPa. The terrain resembled a loose accumulated sandy soil of the kind that would have been generated by the loose spreading action of the blade of a bulldozer or motor grader. The rubber grousers in this case were made of a rubber of spring hardness of $H_s = 62$. As depicted in Figure 4.6, the model-track-plate was equipped with a set of equilateral trapezoidal grousers of trim angle $\alpha = \pi/6$ rad, base length $L = 2, 3, 4$ and 5 cm, and of height $H = 6.5$ cm arranged to a pitch $G_p = 14.6$ cm.

In one series of studies, the effects of the base length of rubber grouser on the terrain-track system constants were investigated. Figure 4.7 presents the results of such a series of model-track-plate loading tests. From the diagram, it can be seen that the magnitude of the static amount of sinkage s_0 for a given contact pressure p increases with a decrease in the base length L.

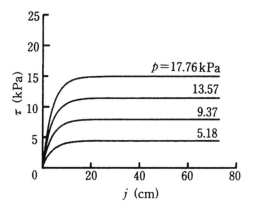

Figure 4.8. Relationship between shear resistance τ and amount of slippage j ($L=4$ cm).

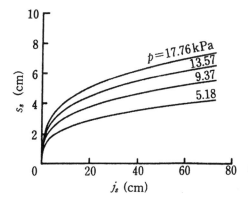

Figure 4.9. Relationship between amount of slip sinkage S_s and amount of slippage j_s ($L=4$ cm).

For a corresponding model-track-plate traction test, the measured relationship between the shear resistance τ and the amount of slippage j for a parameter value of $L=4$ cm is given in Figure 4.8. Similarly, the relationship between the amount of slip sinkage s_s and amount of slippage j_s for $L=4$ cm is shown in Figure 4.9. The τ–j relations shown belong to a type of exponential function as discussed in relation to the previous Eq. (4.5) where m_c equals zero, and the m_f and a values vary with L.

Figure 4.10 shows a plot of the relationships between m_f, a and base length L. From this, it is noted that m_f takes a maximum value at $L=3 \sim 4$ cm and a takes a minimum value at $L=3$ cm. On the other hand, Figure 4.11 shows a plot of the relationships between amount of slip sinkage s_s and base length L. Under these conditions it can be seen that s_s takes a maximum value at $L=4$ cm.

Table 4.2 presents a summary of all the various terrain-track system constants obtained from the above experiments.

4.3.3   Studies on pavement road surfaces

In this section, some experimental results for a track-plate traction test are presented. In this case results were obtained for a model-track-plate equipped with rectangular and

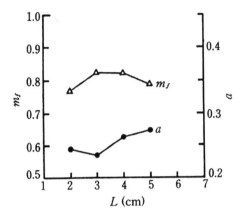

Figure 4.10. Relationship between constants m_f, a and base length L.

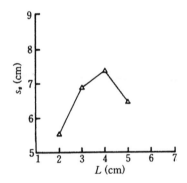

Figure 4.11. Relationship between amount of slip sinkage S_s and base length L ($j_s = 73$ cm, $p = 17.76$ kPa).

Table 4.2. Terrain-track system constants (decomposed granite soil).

L (cm)	2	3	4	5
k_1 (N/cm$^{n_1+2}$)	8.526	11.76	17.05	21.56
n_1	0.866	1.188	1.082	1.202
m_c (kPa)	0	0	0	0
m_f	0.769	0.824	0.822	0.788
a	0.244	0.233	0.260	0.272
c_0 (cm$^{2c_1-c_2+1}$/N^{c_1})	1.588	0.996	0.566	0.448
c_1	0.075	0.264	0.453	0.436
c_2	0.240	0.274	0.295	0.330

equilateral trapezoidal rubber grousers acting on concrete and asphalt pavement road surfaces. The roughness of the asphalt pavement road surface was greater than that of the concrete pavement, and the traction tests were executed under air-dried conditions.

In both the cases, the pavement road surface did not deform during the track plate loading test and hence the terrain-track system constants k_1, n_1 and c_0, c_1, c_2 must be

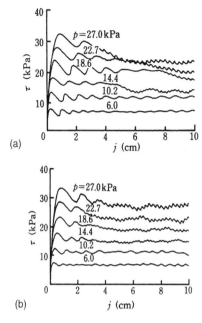

(a)

(b)

Figure 4.12. (a) Relationship between shear resistance τ and amount of slippage j on concrete pavement road (equivalent trapezoidal grouser, $L = 3$ cm). (b) Relationship between shear resistance τ and amount of slippage j on asphalt pavement road (equivalent trapezoidal grouser, $L = 3$ cm).

zero. The rubber grousers used in the trials were made of rubber of a spring hardness of $H_s = 62$. Two kinds of rectangular grouser of height $H = 6.5$ cm and base length of $L = 3$ and 5 cm respectively, and four kinds of equilateral trapezoidal grousers of trim angle $\pi/6$ rad, height $H = 6.5$ cm and base length $L = 2, 3, 4$ and 5 cm respectively were employed. The model-track-plate of width of 25 cm was fitted with a set of five rectangular or equilateral trapezoidal grousers arranged to a pitch $G_p = 14.6$ cm and $G_p/H = 2.25$.

Under these conditions, several aspects of the influence of the type of grousers and the base length of the grousers on the terrain-track system constants have been studied. Figure 4.12(a) graphs the discovered relationship that exists between the shear resistance τ and the amount of slippage j of the equilateral trapezoidal grousers.

The grousers had a base length of $L = 3$ cm and were tested on a concrete pavement road under various contact pressures p. Figure 4.12(b) shows the same data obtained for tests on an asphalt pavement.

From these tests, it can be seen that both of the τ–j relations belong to the Hump type category as discussed previously in relation to Eq. (4.6). Given this hump-type relation, the terrain-track system constants f_s, f_m and j_m can thence be determined. Figure 4.13 shows a plot of the relationship between f_s and base length L for several combinations of types of grousers operating on both concrete and asphalt pavement road surfaces.

Figure 4.14 gives the relationships between f_m and L for the same combinations of grouser and road pavement type. Since equilateral trapezoidal grousers are structurally stronger than the rectangular ones, the constants f_s and f_m for the equilateral trapezoidal grousers yield higher values than those for the rectangular grousers.

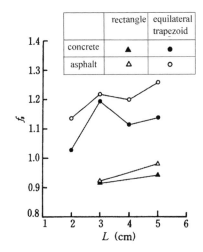

Figure 4.13. Relationship between constant f_s and base length L.

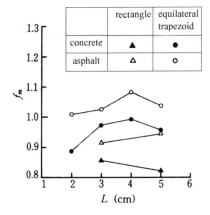

Figure 4.14. Relationship between constant f_m and base length L.

While both the values of f_s and f_m show some characteristics changes with base length L, the values for asphalt pavement are generally larger than those for concrete pavements.

Table 4.3 summarises all the terrain-track system constants obtained from this particular series of trials.

4.3.4 Scale effects and the model-track-plate test

In discussions to date, a variety of model tests have been introduced which have sought to clarify the mechanics of the real-world interaction between a soil and a piece of construction machinery through the use of models. However it is well recognised in engineering circles that in many situations it is difficult to apply the results of a model test directly to a full scale problem because of size-effect problems. To investigate the nature of scale effects, a number of geometrically similar model tests were tested to estimate, for example, the

Table 4.3. Terrain-track system constants (road pavement).

Road pavement	Type of grouser	L (cm)	f_s	f_m	j_m (mm)
Concrete	Rectangular	3	0.918	0.856	6.17
		5	0.944	0.823	5.76
	Equilateral trapezoid	2	1.028	0.887	8.87
		3	1.193	0.972	4.84
		4	1.113	0.992	6.25
		5	1.138	0.959	3.75
Asphalt	Rectangular	3	0.920	0.916	10.87
		5	0.980	0.949	4.89
	Equilateral trapezoid	2	1.134	1.010	6.62
		3	1.217	1.026	5.61
		4	1.200	1.081	6.86
		5	1.259	1.037	3.93

Table 4.4. Modified list of variables.

Variable	Symbol	Basic dimensions
Lengths	λ, λ_a	L
Forces	R, R_1	F
Soil properties	α	$F^a L^b$
	α_i	$F^{a_i} L^{b_i}$

bulldozing resistance of the blade of a bulldozer. From such tests, the actual behaviour of a blade can be estimated from a model by application of the principles of dimensional analysis [9,10].

Muro and Kawahara [11] confirmed the existence of a size effect in the estimation of actual tractive resistances and amounts of sinkage through use of the model-track-plate traction test. The study was done to investigate the trafficability of a tracked vehicle running on a soft terrain. In the study, the researchers considered a distorted model condition (see later explanation) in which the same soil was used for the model test as was the actual terrain upon which the full scale construction machinery was operating. The researchers developed a quantitative approach to estimate the actual tractive resistance and amount of sinkage as a function of a ratio of geometrical similarity by use of dimensional analysis consideration and through use of the well known Buckingham-Pi principles of analysis which uses a number of non-dimensional parameters or groups called Π terms.

Schafer et al. [12] took the general variables as shown in Table 4.4 to cover the similarity problem of soil-machine system constants. They assumed that the variables of acceleration and so on that included a dimension of time could be neglected in the field of terramechanics since agricultural and construction machinery move comparatively slowly. From the list of selected variables, they developed the following non-dimensional Π terms;

$$\Pi_1 = \frac{R}{\alpha^{(1/a)} \lambda^{(b/a)}} \tag{4.12}$$

$$\Pi_l = \frac{R_1}{R} \tag{4.13}$$

$$\Pi_q = \frac{\lambda_q}{\lambda} \tag{4.14}$$

$$\Pi_{\alpha_i} = \frac{\alpha_i}{\alpha^{(a_i/a)}\lambda^{b_i - (ba_i/a)}} \tag{4.15}$$

where Π_l is a term which includes the factor R which should be estimated, Π_l is a term comprised of a ratio between several factors, Π_q is likewise a non-dimensional term comprised of the ratio between several dimensions, Π_{α_i} is a term that includes the soil constants. The factor n is a ratio of similarity e.g. $n_R = R/R_m$, $n_{R1} = R_1/R_{1m}$, $n_\lambda = \lambda/\lambda_m$, $n_\alpha = \alpha/\alpha_m$, $n_{\lambda q} = \lambda_q/\lambda_{qm}$. The subscript m expresses the model relationship. Under these conditions, the design relationships which need to be satisfied for similitude are as follows:

① $\quad n_R = n_\alpha^{(1/a)} n_\lambda^{(b/a)}$ $\hfill (4.16)$

② $\quad n_{R1} = n_R$ $\hfill (4.17)$

③ $\quad n_{\lambda q} = n_\lambda$ $\hfill (4.18)$

④ $\quad n_{a_i} = n_\alpha^{(a_i/a)} n_\lambda^{b_i - (ba_i/a)}$ $\hfill (4.19)$

For the above $l + q + I + 1$ conditional equations, the number of unknown factors are $l + q + I + 3$ in which the two values of n_λ and n_α can be given arbitrarily. Here, the condition which is usually considered is $n_\alpha = n_{\alpha_i} = 1$, if α is the same as α_i. It is noted, however, that it is additionally necessary to satisfy either of the next conditions to establish condition ④.

⑤ $\quad \alpha_i$ is a non-dimensional quantity e.g. $a_i = b_i = 0$.
⑥ $\quad \alpha$ has the same dimension as that of α_i.
⑦ $\quad$ The dimensions of α and α_i satisfy the equation $b_i = ba_i/a$.

As the condition ④ is distorted when these conditions are not satisfied, a distorted model system should be considered. A coefficient of distortion β_i for the soil constants can be defined as follows:

$$\beta_i = \frac{\prod \alpha_i m}{\prod \alpha_i} = n_\lambda^{b_i - (ba_i/a)} = n_\lambda^{s_i} \tag{4.20}$$

$$s_i = b_i - \frac{ba_i}{a} \tag{4.21}$$

That is, β_i can be expressed as a function of the ratio of similarity n_λ.

Moreover, for the distorted model system, it is necessary to define the following coefficient of estimation δ of Π terms. This collection should include the force which should be estimated:

$$\delta = \frac{\Pi_1}{\Pi_{1m}} = \frac{f(\Pi_2, \Pi_3, \ldots, \Pi_c)}{f(\Pi_{2m}, \Pi_{3m}, \ldots, \Pi_{cm})} \tag{4.22}$$

When Π_1 and the distorted Π_{α_i} are expressed by a power function, the factor δ can then be presented as follows:

$$\delta = n\lambda^s$$
$$s = -\sum e_i s_i \qquad\qquad (4.23)$$

where e_i is the index of Π_{α_i} in the relation between Π_1 and Π_{α_i}.

These results show that, when a model test is carried out by use of the same soil as that of the actual terrain, an estimation of the actual behaviour of a prototype from a model test result is possible even if the soil constants belong to the distorted system. δ is a function of size ratio n_λ only, and the index s is constant for a given soil-machine system.

As an example of the use of a traction test employing a model-track-plate to investigate the size effect, several dimensional analytical results are presented here.

The data given are for the maximum tractive resistance F_{max} and the amount of sinkage z_m of a standard T shaped model-track-plate running on a weak silty loam terrain. The silty loam terrain was prepared by remolding the loam uniformly and then adjusting it to water contents of 30, 35 and 40%. The standard T shaped model-track-plate employed is shown schematically in Figure 4.15. In total, twelve different kinds of model-track-plate with grouser pitch–height ratio of $G_p/H = 1.5$, 3.0 and 4.5 and with size ratios of $n_\lambda = 1$, 2, 4 and 8, as shown in Table 4.5, were employed. In these tests, the contact pressure p was alternately selected as 0.98, 2.94, 4.90, 6.86 and 9.80 kPa.

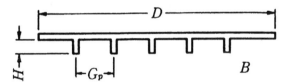

Figure 4.15. Standard T shaped track model plate.

Table 4.5. Shape and dimension of track model plate.

Track model	Track length D (cm)	Track width B (cm)	Grouser height H (cm)	Grouser pitch G_p (cm)	G_p/H
I	29.0	5.0	1.0	4.5	4.5
	29.0	5.0	1.5	4.5	3.0
	29.0	5.0	3.0	4.5	1.5
II	58.0	10.0	2.0	9.0	4.5
	58.0	10.0	3.0	9.0	3.0
	58.0	10.0	6.0	9.0	1.5
III	116.0	20.0	4.0	18.0	4.5
	116.0	20.0	6.0	18.0	3.0
	116.0	20.0	12.0	18.0	1.5
IV	232.0	40.0	8.0	36.0	4.5
	232.0	40.0	12.0	36.0	3.0
	232.0	40.0	24.0	36.0	1.5

In these tests, the width B (cm), the length D (cm), the grouser pitch G_p (cm) and the height H (cm) of the model-track-plate were selected as the variables in the size dimension whilst the maximum tractive resistance F_{max}(N) and the load W(N) were selected as the variables in the force dimension. The undrained shear resistance c_u (kPa) (FL^{-1}) and the weight per unit volume γ (kN/m^3) (FL^{-3}) were selected as the variables in the soil constants.

For a value of F_{max} which needs to be estimated, the experimental results can be arranged in the following Π terms:

$$\Pi_1 = \frac{F_{max}}{c_u B^2} \qquad \Pi_2 = \frac{W}{c_u BD} = \frac{p}{c_u}$$

$$\Pi_3 = \frac{G_p}{H} \qquad \Pi_4 = \frac{\gamma B}{c_u} \tag{4.24}$$

For the independent variable Π_1 and the dependent variables Π_2, Π_3 and Π_4, all the experimental data may be analysed through use of multi-regression analysis. In relation to the results, Π_1 can be expressed as the product of Π_2, Π_3 and Π_4 as shown in the next equation:

$$\Pi_1 = 13.07\, \Pi_2^{0.496}\, \Pi_3^{-0.586}\, \Pi_4^{-0.333} \tag{4.25}$$

Substituting the previous Eq. (4.24) and the value $\gamma = 18.2\,\mathrm{kN/m^3}$ into above equation, F_{max} can be approximated as follows:

$$F_{max} = 4.97 c_u^{0.837} p^{0.496} B^{1.667} \left(\frac{G_p}{H}\right)^{-0.586} \tag{4.26}$$

When the soil constants of model test are the same as those of prototype, $n_{cu} = n_\gamma = 1$, and $s_i = -1, s = -0.333$ for the indices $a = 1, b = -2, a_i = 1, b_i = -3$. Then the coefficient of distortion β and the coefficient of estimation δ can be formulated as:

$$\beta = \frac{\Pi_{4m}}{\Pi_4} = n_B^{-1} \tag{4.27}$$

$$\delta = \frac{\Pi_1}{\Pi_{1m}} = n_B^{-0.333} \tag{4.28}$$

Thence, substituting the value of Π_1 given in Eq. (4.24) into Eq. (4.28), the ratio of $(F_{max})_p$ of the prototype to $(F_{max})_m$ of the model can be expressed as:

$$\frac{(F_{max})_p}{(F_{max})_m} = n_B^{1.667} \tag{4.29}$$

Where the value of the index is greater than, say, 2.0 then there is no size effect. In this case, the reason for the occurrence of the size effect is considered to be the fact that the length of the slip line in front of the top grouser and the side area between grousers does not increase linearly with increments in grouser height H because the amount of sinkage of the model-track-plate is not in proportion to the grouser height H.

Next, we can define the amount of sinkage z_m as the amount of sinkage measured at the center of the model-track-plate when the tractive resistance takes a maximum value F_{max}.

For a value of z_m, which needs to be estimated, the experimental results can be arranged in the following Π terms:

$$\Pi_1' = \frac{z_m}{H} \qquad\qquad \Pi_2 = \frac{p}{c_u}$$

$$\Pi_3 = \frac{G_p}{H} \qquad\qquad \Pi_4' = \frac{\gamma H}{c_u} \tag{4.30}$$

Additionally, to satisfy the design condition ④ in this experiment, it is necessary to control the soil constants so that they are uniform in the direction of depth. Taking the results of a multi-regression analysis for all the experimental data, Π_1' can be expressed as the product of Π_2, Π_3 and Π_4' as shown in the next equation:

$$\Pi_1' = 0.705\, \Pi_2^{0.559}\, \Pi_3^{0.312}\, \Pi_4'^{-0.046} \tag{4.31}$$

Substituting the previous Eq. (4.30) into above equation, z_m can be approximated as follows:

$$z_m = 0.617 c_u^{-0.513} p^{0.559} H^{0.954} \left(\frac{G_p}{H}\right)^{0.312} \tag{4.32}$$

As the term Π_4' is the distorted Π term, the coefficient of distortion β and the coefficient of estimation δ can be expressed, in a similar way to the case of F_{max}, as follows:

$$\beta = \frac{\Pi_{4m}'}{\Pi_4'} = n_H^{-1} \tag{4.33}$$

$$\delta = \frac{\Pi_1'}{\Pi_{1m}'} = n_H^{-0.046} \tag{4.34}$$

Thence, substituting the value of Π_1 given in the previous Eq. (4.30) into the above Eq. (4.34), the amount of sinkage of a prototype $(z_m)_p$ can be expressed by use of that of the model $(z_m)_m$ as follows:

$$\frac{(z_m)_p}{(z_m)_m} = n_H^{0.954} \tag{4.35}$$

Since the index n_H is less than 1.0, it is evident that the relative amount of sinkage Π_1' of the model is larger than that of the prototype. This occurs because, for the small size of the model-track-plate, the relative amount of sinkage Π_1' becomes large due to the short average moving distance of the soil when the sandwiched soil material between the grousers is displaced sideways, while, for the large size of model-track-plate, the amount of sinkage is restrained due to difficulty in the lateral or sideways movement of the sandwiched soil elements between the grousers.

In summary of all the above, it is clearly evident that there are some size effects on both the values of F_{max} and z_m in the traction test of the model-track-plate.

Next, let us look at some test results that study several of the size effects that operate in the tractive performance of a model-track-plate equipped with eight equilateral trapezoidal

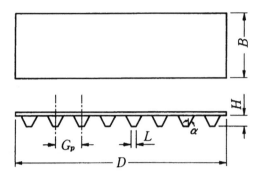

Figure 4.16. Equivalent trapezoidal track model plate.

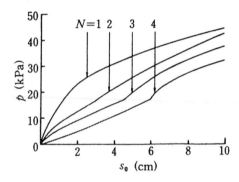

Figure 4.17. Relationship between contact pressure p and static amount of sinkage s_0 for various size ratios N.

rubber grousers. The cross section of the test plate is as depicted in Figure 4.16. The tests were carried out on an unsaturated weak terrain. Four kinds of geometrically similar model-track-plates having a size ratio of $N = 1$, 2, 3 and 4 were utilised in the plate loading and traction test.

The dimensions of the model-track-plate for $N = 1$ are as follows: contact length $D = 41$ cm, width $B = 12.5$ cm, grouser height $H = 1.5$ cm, grouser pitch $G_p = 4.5$ cm, trim angle $\alpha = \pi/6$ rad and base length of grouser $L = 0.5$ cm. The grouser pitch–height ratio for each size ratio was set equal to 3.0 to maximize tractive resistance. The unsaturated weak terrain was prepared by remolding a silty loam to a water content of 30%. This material was then filled into a large soil bin ($540 \times 150 \times 60$ cm) until a depth of 40 cm was achieved. The measured cone index of the terrain was 31 kPa and the degree of saturation 96.9%.

Figure 4.17 shows the results of carrying out a plate loading test on a model-track-plate. The values of s_0 (cm) for a constant value of contact pressure p (kPa) increased with increment of N because the range of influence of the pressure bulb extends due to the increasing total load applied to the model-track-plate. Figure 4.18 shows the relations between the terrain-track system constants $k_1 (N/cm^{n_1+1})$, n_1, $k_2 (N/cm^{n_2+1})$, n_2 given in previous Eqs. (4.1), (4.2) and the size ratio N. The factors k_1, k_2 decrease with increasing N, but n_1, n_2 increase with N.

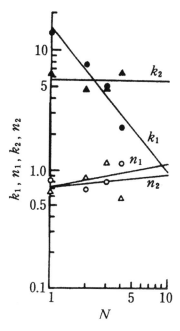

Figure 4.18. Relationship between constants k_1, n_1, n_2, and size ratio N.

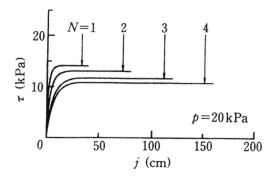

Figure 4.19. Relationship between shear resistance of soil τ and amount of slippage j for various size ratios N.

The resulting, experimentally derived, equations are as follows:

$$k_1 = 1.50 \times 10N^{-1.192} \tag{4.36}$$

$$k_2 = 5.69N^{-0.015} \tag{4.37}$$

$$n_1 = 7.30 \times 10^{-1}N^{0.183} \tag{4.38}$$

$$n_2 = 7.30 \times 10^{-1}N^{0.098} \tag{4.39}$$

To illustrate these concepts, Figure 4.19 shows the results of a model-track-plate traction test i.e. the relationships between shear resistance τ (kPa) and amount of slippage j (cm), for various values of N. The value of τ decreases with increasing values of N, because the

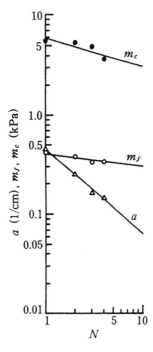

Figure 4.20. Relationship between constants m_c, m_f, a and size ratio N.

bulldozing resistance that acts in front of the model-track-plate tends to decrease due to a decreasing amount of sinkage of the front part of the model-track-plate accompanied by an increasing angle of inclination of the plate.

Figure 4.20 shows the relationships between the terrain-track system constants m_c (kPa), m_f, and a (1/cm) given in the previous Eq. (4.5) and the size ratio N. Each value decreases with increasing N. The experimental equations that best fit this data are:

$$m_c = 6.04N^{-0.282} \tag{4.40}$$

$$m_f = 4.13 \times 10^{-1} N^{-0.137} \tag{4.41}$$

$$a = 4.35 \times 10^{-1} N^{-0.824} \tag{4.42}$$

Figure 4.21 shows, as an illustrative study, the relationship between the amount of slip sinkage s_s (cm) of a model-track-plate and the amount of slippage j_s (cm) for various values of N. The factor s_s increases with increasing values of contact pressure p and amounts of slippage j_s because of an increasing bulldozing action and volume change in the soil due to the dilatancy phenomenon that occurs during shear action. Additionally it is noted that s_s increases with increasing values of N because the influence zone of the pressure bulb penetrates more deeply.

Figure 4.22 shows the relations between the terrain-track system constants c_0, c_1, and c_2 given in the previous Eq. (4.8) and the size ratio N.

$$c_0 = 4.19 \times 10^{-2} N^{1.349} \tag{4.43}$$

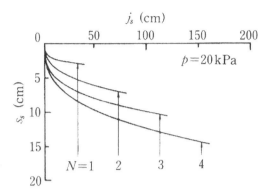

Figure 4.21. Relationship between amount of slip sinkage s_0 and amount of slippage j_s.

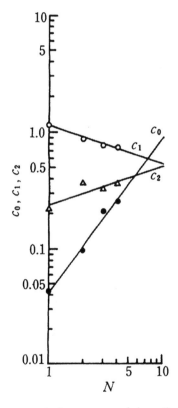

Figure 4.22. Relationship between constants c_0, c_1, c_2 and size ratio N.

$$c_1 = 1.15N^{-0.324} \tag{4.44}$$

$$c_2 = 2.40 \times 10^{-1}N^{0.331} \tag{4.45}$$

As to these results, both the values of c_0, c_2 increase with increases in N. In contrast, though, the value of c_1 decreases with N.

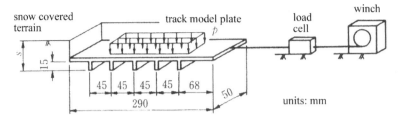

Figure 4.23. Traction test of standard T shaped track model plate.

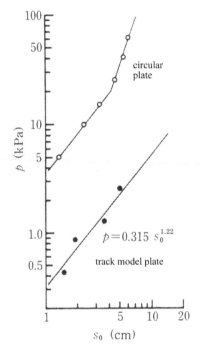

Figure 4.24. Relationship between contact pressure p and static amount of sinkage s_0 for circular plate and track model plate.

4.3.5 Snow covered terrain

In this section we present some experimental test results relative to the undertaking of a loading test and a traction test on snow. In this case, the apparatus consisted of a model-track-plate equipped with standard T shaped grousers. The terrain was a snow covered one composed of newly fallen winter snow at an outside temperature of 0 °C.

The depth of the snow was 20 cm. The average density of the newly fallen snow ρ_0 was 0.228 g/cm^3 and the water content of the wet snow was 1.02%. Both a circular plate of diameter of 7.0 cm and a model-track-plate of contact pressure of $D = 29$ cm, width $B = 5$ cm, grouser height $H = 1.5$ cm, and grouser pitch of $G_p = 4.5$ cm were used in the loading trials. A cross section of the apparatus employed is shown in Figure 4.23.

Figure 4.24 gives the results of the loading tests for the circular plate and the model-track-plate respectively. It can be seen in this graph that, for a constant contact pressure,

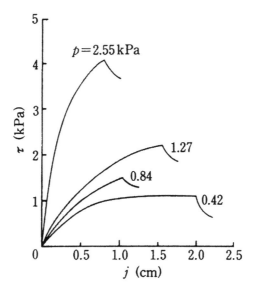

Figure 4.25. Relationship between shear resistance of snow τ and amount of slippage j for various contact pressures.

the amount of sinkage of the model-track-plate becomes larger than that of the circular plate. This is presumed to be due to the penetration action of the grouser itself.

Following the straight loading test, a series of traction tests on the model-track-plate were carried out at various levels of constant pressure $p = 0.42, 0.84, 1.27$ and $2.55\,\text{kPa}$. To assure a traction-speed of $3.7\,\text{mm/s}$, the model-track-plate was pulled by a wire and winch.

Figure 4.25 shows the results for the shear resistance τ (kPa) of the snowy terrain and the amount of slippage j (cm) plotted for various levels of contact pressure p (kPa). As is very evident from the graphs, each curve in this series follows a Hump type relationship in which the shear resistance τ decreases rapidly after a peak. This is due to a brittle failure within the snow material after it passes a maximum value. For each dynamic slip sinkage relation in the traction test, it is observed that the total amount of sinkage s (i.e. the sum of the static amount of sinkage s_0 and the dynamic amount of slip sinkage s_s) increases rapidly after the slippage j_s attains a value of around $1 \sim 2$ cm. This rapid increase happens as a result of the dramatic collapse of the snow material. After this, the amount of slip sinkage increased intermittently with increments in applied load.

From the above series of test, the terrain-track system constants for a representative snowy terrain can be summarized as having the following values: $k_2 = 0.315$, $n_2 = 1.220$, $k_4 = 32.34$, $n_4 = 0.862$, $f_s = 1.86 \pm 0.35$, $f_m = 0.01$, $j_m = 1 \sim 2$ cm and $c_0 = 0.685$, $c_1 = 0.694$ and $c_2 = 0.476$.

4.4   SUMMARY

In this chapter, we have set the scene for a general study of tracked vehicle systems. Because of the presence of grousers and the existence of the phenomenon of slip-sinkage, the behaviour of tracked running-gear systems is totally different from that of cylindrical drums and wheels. A new modelling concept is therefore required to predict behaviour. In this chapter we have proposed the development of a prediction method based upon a set of experimentally derived parameters. The parameters are developed from small scale, shaped-plate, traction tests. They are scaled up to real machinery sizes using the theory of dimensional analysis and similitude.

REFERENCES

1. Tamura, Y. & Kaminishi, M. (1980). Improvement of the Tractive Performance with New Track System (Multi-rollers). *Komatsu Technical Review, Vol. 26, No. 3*, pp. 1–7. (In Japanese).
2. Akai, K. (1986). *Soil Mechanics*. pp. 86–112. Asakura Press. (In Japanese).
3. Janosi, Z. & Hanamoto, B. (1961). The Analytical Determination of Drawbar Pull as a Function of Slip For Tracked Vehicles. *Proc. 1st Int. Conf. on Terrain-Vehicle Systems*. Torio.
4. Oida, A. (1975). Study on equation of shear stress–displacement curves – on Kacigin-Guskov equation – *J. of Agricultural Machinery, Vol. 37, No. 1*, pp. 20–25. (In Japanese).
5. Bekker, M.G. (1960). *Off-the-road Locomotion*. pp. 58–66. The University of Michigan Press.
6. Yamaguchi, H. (1984). *Soil Mechanics*. pp. 141–196. Gihoudou Press. (In Japanese).
7. Hata, S. & Hosoi, T. (1981). On the effect of lug pitch upon the tractive effort about track-laying vehicle, *Proc. 7th Int. Conf. ISTVS, Calgary, Canada. Vol. 1*, pp. 255–262.
8. Muro, T., Omoto, K. & M. Futamura. (1988). Traffic performance of a bulldozer running on a weak terrain – Vehicle model test, *J. of JSCE. No.397/VI-9*, pp. 151–157. (In Japanese).
9. Ketterer, B. (1981). Modelluntersuchungen zur prognose von schneid und planierkräften im erdbau. *BMT, 7*, pp. 355–370.
10. Freitag, D.R., Schafer, R.L. & Wismer, R.D. (1970). Similitude Studies of Soil Machine Systems, *Trans. ASAE, Vol. 13, No. 2*, pp. 201–213.
11. Muro, T. & Kawahara, S. (1986). Size effect of Track Performances of Running Gear on Weak Ground. *J. of JSCE, No. 370/III-5*, pp. 105–112. (In Japanese).
12. Schafer, R.L., Reaves, C.A. & Young, D.F. (1969). An Interpretation of Distortion in the Similitude of Certain Soil-Machine Systems. *Trans. ASAE, Vol. 12, No. 1*, pp. 145–149.

EXERCISES

(1) Suppose that three different kinds of standard T shaped steel track of grouser height $H = 3.2$ cm and grouser pitch $G_p = 5.1$, 10.2 and 20.4 cm are standing at rest on a silty loam terrain and that they each are sustaining a contact pressure of $p = 9.8$ kPa. Calculate the static amount of sinkage s_0 for each system by use of the terrain-track system constants data given in Table 4.1. Assume that the value of p_0 in Eq. (4.2) is 1.96 kPa for all the tracks.

(2) Assume that there are three kinds of standard T shaped steel track equipped with grousers at a pitch $G_p = 5.1$, 10.2 and 20.4 cm. Suppose also that the track has a length of 70 cm and a width of 20 cm and that the contact pressure on the track is 7.84 kPa. For this arrangement, calculate the tractive resistance T when the track is pulled until a slip value of 5 cm is attained. Use the terrain-track system constants of Table 4.1.

(3) Suppose that a standard T shaped steel track having three kinds of grouser pitch G_p is slipping under a contact pressure of 7.84 kPa. Calculate the amount of slip sinkage s_s of the track when the amount of slippage j_s reaches 7 cm. Assume the terrain-track system constants are as given in Table 4.1.

(4) Imagine that a set of equilateral trapezoidal shaped rubber tracks equipped with four kinds of grouser of base length $L = 2$, 3, 4 and 5, as shown in Figure 4.6, are in a stationary state on a loose accumulated decomposed granite sandy terrain. The contact pressure is 29.4 kPa. Calculate the static amount of sinkage s_0 of each system by use of the terrain-track system constants of Table 4.2.

(5) Equilateral trapezoidal shaped rubber tracks equipped with four kinds of grouser of base length $L = 2$, 3, 4 and 5 are pulled on a loose sandy terrain. The track has a length of L of 50 cm and a width of 20 cm and the contact pressure is 34.3 kPa. Calculate the tractive force T when the amount of slippage reaches 6 cm, through use of the terrain-track system constants of Table 4.2.

(6) Four kinds of equilateral trapezoidal shaped rubber track equipped with grousers having different base length L are slipping under a contact pressure $p = 53.9$ kPa. Calculate the amount of slip sinkage s_s when the amount of slippage j_s reaches 50 cm. The terrain-track system constants are as shown in Table 4.2.

(7) An equilateral trapezoidal shaped rubber track equipped with a grouser of base length $L = 4$ cm is pulled on a concrete pavement. The track has a length of 70 cm, a width of 25 cm and is operating under a contact pressure of 29.4 kPa. Calculate the tractive resistance T when the amount of slippage j reaches 8.5 cm. Assume that the terrain-track system constants are as shown in Table 4.3.

(8) An equilateral trapezoidal shaped rubber track of length 41 cm and width 12.5 cm is operating in a stationary state on a silty loam terrain. Calculate the static amount of sinkage for a contact pressure p of 25.5 kPa. How does the amount of sinkage s_0 vary for same contact pressure when the size ratio N increases from 2, to 3 and then to 4? Assume that the height of the grouser H is 1.5 cm for a size ratio $N = 1$, and that the coefficients of sinkage k_1, k_2 and the indices of sinkage n_1, n_2 are as given in Eqs. (4.36) ~ (4.39).

(9) When the size ratio N of the track given in the previous problem (8) increases through $N = 1$, 2, 3 and 4, calculate the variation in the shear resistance τ of the soil at the amount of slippage $j = 10$ N (cm). Assume that the terrain-track system constants are as given in Eqs. (4.40) ~ (4.42). When the size ratio N of the track given in the previous problem (8) increases through $N = 1$, 2, 3 and 4, calculate the variation in the amount of slip sinkage s_s of the tracks at an amount of slippage $j_s = 5N$ (cm). Assume the terrain-track system constants are as given in Eqs. (4.43) ~ (4.45).

Chapter 5

Land Locomotion Mechanics for a Rigid-Track Vehicle

5.1   REST STATE ANALYSIS

5.1.1   Bearing capacity of a terrain

When a rigid tracked vehicle of weight W, track belt contact length D, track width B, and amount of eccentricity of center of gravity of vehicle e is in a stationary condition on a hard terrain as shown in Figure 5.1, a ground reaction P acts on the track belt at the position of the amount of eccentricity $e_0 D$ and the distribution of the contact pressure takes on the shape of a trapezoid within the bounds $|e_i| \leq 1/6$. In this case, the amount of sinkage of the track belt is negligibly small for a hard terrain. The contact pressures at the front and rear ends of the track belt p_f and p_r can be expressed as follows providing that we assume that the hard terrain is a pure linearly-elastic material:

$$p_f = p_m(1 - 6e_0) \tag{5.1}$$

$$p_r = p_m(1 + 6e_0) \tag{5.2}$$

$$p_m = \frac{W}{2BD} \tag{5.3}$$

where p_m is the average contact pressure. On the other hand, the distribution of the contact pressure becomes a triangle to the rear-side for $e_0 > 1/6$ and to the front side for $e_0 < -1/6$.

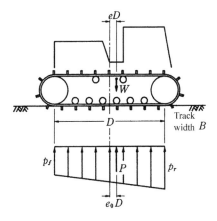

Figure 5.1. Contact pressure distribution under track belt on hard terrain (at rest).

For these static conditions, the bearing capacity Q_c of the terrain can be expressed, following Meyerhof [1], as:

$$Q_c = 2BD_e \left(cN_{cq}\lambda_{cq} + \frac{\gamma BN_{rq}\lambda_r}{2} \right) \tag{5.4}$$

where c is the cohesion, N_{cq}, $N_{\gamma q}$ is the coefficient of bearing capacity, λ_{cq}, λ_γ is the shape coefficient, and $D_e = D(1 - 2e_0)$ is the effective contact length of the track belt which considers the effective loading area. However, when a tractive effort or a braking force acts on the vehicle, it is necessary to modify the above equation for inclined load because of the inclined and eccentric loads that are applied to the track belt.

When e_0 equals zero, Q_c can be also expressed following Terzaghi [2] as:

$$Q_c = 2BD \left\{ \left(1 + \frac{0.3B}{D} \right) cN_c + \left(0.5 - \frac{0.1B}{D} \right) B\gamma N_\gamma \right\} \tag{5.5}$$

where N_c and N_γ are the coefficients of bearing capacity.

5.1.2   Distribution of contact pressures and amounts of sinkage

When a uniform distribution of contact pressure p is applied through a rigid track belt of width B, as shown schematically in Figure 5.2, several equipotential lines of principal stress [3] can be drawn for the region under the track. This can be done through use Boussinesq's elastic solution for a uniformly distributed strip load.

The maximum principal stress σ_1 acting at an arbitrary point A on the equipotential line of the principal stress operates in a direction that bisects the central angle 2ε as shown in this diagram. Under the same conditions the minimum principal stress σ_3 acts perpendicular to σ_1. The values of σ_1 and σ_3 are thence as follows:

$$\sigma_1 = (2\varepsilon + \sin 2\varepsilon)\frac{p}{\pi} \tag{5.6}$$

$$\sigma_3 = (2\varepsilon - \sin 2\varepsilon)\frac{p}{\pi} \tag{5.7}$$

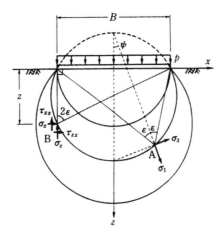

Figure 5.2. Equi-principal stress lines under track belt.

Further, the horizontal stress σ_x, the vertical stress σ_z and the shear stress τ_{xz} that acting at the point A are:

$$\sigma_x = (2\varepsilon - \sin 2\varepsilon \cdot \cos 2\psi)\frac{p}{\pi} \tag{5.8}$$

$$\sigma_z = (2\varepsilon + \sin 2\varepsilon \cdot \cos 2\psi)\frac{p}{\pi} \tag{5.9}$$

$$\tau_{xz} = \sin 2\varepsilon \cdot \sin 2\psi \cdot \frac{p}{\pi} \tag{5.10}$$

Then, the vertical contact pressure σ_z acting at an arbitrary point in the terrain can be calculated using Eq. (5.9).

On the other hand, the horizontal stress σ_x, the vertical stress σ_z and the shear stress τ_{xz} acting at a point B where a vertical line through the edge of the track belt intersects the equipotential line of principal stress can be expressed as:

$$\sigma_x = (2\varepsilon - \sin 2\varepsilon \cdot \cos 2\varepsilon)\frac{p}{\pi} \tag{5.11}$$

$$\sigma_z = (2\varepsilon + \sin 2\varepsilon \cdot \cos 2\varepsilon)\frac{p}{\pi} \tag{5.12}$$

$$\tau_{xz} = \sin^2 2\varepsilon \cdot \frac{p}{\pi} \tag{5.13}$$

Then, the horizontal stress σ_x acting on the side part of grouser of the track belt can be calculated by use of Eq. (5.11).

The normal resultant force F_x acting on the side part of grouser over the whole range of the contact length of the track belt can be calculated by integrating σ_x from zero to the grouser height H as follows:

$$2\varepsilon = \cot^{-1}\left(\frac{z}{B}\right)$$

$$\sin 2\varepsilon = \frac{B}{\sqrt{z^2 + B^2}}, \qquad \cos 2\varepsilon = \frac{z}{\sqrt{z^2 + B^2}}$$

therefore,

$$F_x = D\int_0^H \sigma_x\, dz = \frac{Dp}{\pi}\int_0^H \left\{\cot^{-1}\left(\frac{z}{B}\right) - \frac{Bz}{z^2 + B^2}\right\} dz$$

$$= \frac{DpH}{\pi} \cdot \cot^{-1}\left(\frac{H}{B}\right) \tag{5.14}$$

When a rigid-track vehicle is standing stationary on a soft terrain, the rigid track belt will be generally inclined as a result of an amount of sinkage at the front-idler s_f and an amount of sinkage at the rear sprocket s_r. The angle of inclination of the vehicle θ_{t0} also varies according to the amount of eccentricity e of the center of gravity of the vehicle. As shown in Figure 5.3(a ~ e), the main part of the track belt $\overline{BC}$ can take up five different distinctive positions as a function of the position of the eccentricity e_0 relative to that of the ground reaction P. In these cases, the amounts of sinkage s_{f0} and s_{r0} are measured in directions perpendicular to the normal direction of the main part of the track belt. The distances are taken from the ground surface to the tip of the grouser.

The amount of sinkage $s_0(X)$ at a distance X from a point B on the front edge of the main part of the track belt can be expressed by the following linear function.

$$s_0(X) = s_{f0} + (s_{r0} - s_{f0})\frac{X}{D} \tag{5.15}$$

where D is the contact length of the rigid track belt.

Of necessity, the distribution of the contact pressure under the rigid track belt cannot be expressed as a linear function. It should be curved in general as shown in the following equations:

For $0 \leqq s_0(X) \leqq H$

$$p_0(X) = k_1 \{s_0(X)\}^{n_1} \tag{5.16}$$

For $s_0(X) > H$

$$p_0(X) = p_0 + k_2 \{s_0(X) - H\}^{n_2} \tag{5.17}$$

The contact pressure p_{f0}, the amount of sinkage s_{f0} at the contact point B between the front-idler and the main part of the rigid track belt, the contact pressure p_{r0} the amount of sinkage s_{r0} at the contact point C between the rear sprocket and the main part of the rigid track belt can be calculated from the appropriate force and moment equilibrium equations [4] as follows:

(1) *For the case where $s_{f0} \geq H, s_{r0} \geq H$*
Using the symbols used in Figure 5.3(a), the following equations can be established:

$$W \cos \theta_{t0} = 2B \int_0^D p_0(X)\, dX = 2BDp_0 + \frac{2k_2BD}{n_2+1} \times \frac{(s_{r0} - H)^{n_2+1} - (s_{f0} - H)^{n_2+1}}{s_{r0} - s_{f0}} \tag{5.18}$$

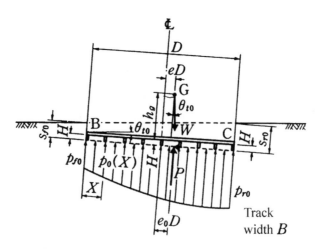

Figure 5.3(a). Distribution of contact pressure $P_i(X)$ and amount of sinkage $S_i(X)$ for $s_{f0} \geq H$, $s_{r0} \geq H$.

$$W \cos \theta_{t0} \cdot \left(\frac{1}{2} + e_0 \right) D = 2B \int_0^D p_0(X) X \, dX$$

$$= \frac{2BD^2}{(s_{r0} - s_{f0})^2} \left[\frac{k_2}{n_2 + 2} \left\{ (s_{r0} - H)^{n_2+2} - (s_{f0} - H)^{n_2+2} \right\} \right.$$

$$- \frac{k_2}{n_2 + 1} (s_{f0} - H) \left\{ (s_{r0} - H)^{n_2+1} - (s_{f0} - H)^{n_2+1} \right\}$$

$$\left. + \frac{1}{2} p_0 (s_{r0}^2 - s_{f0}^2) - p_0 s_{f0} (s_{r0} - s_{f0}) \right] \qquad (5.19)$$

From above equations, s_{f0} and s_{r0} can be calculated as follows:

$$\theta_{t0} = \tan^{-1} \left(\frac{s_{r0} - s_{f0}}{D} \right) \qquad (5.20)$$

$$e_0 = e + \frac{h_g + H}{D} \tan \theta_{t0} \qquad (5.21)$$

$$p_{f0} = p_0 + k_2 (s_{f0} - H)^{n_2} \qquad (5.22)$$

$$p_{r0} = p_0 + k_2 (s_{r0} - H)^{n_2} \qquad (5.23)$$

(2)   *For the case where* $0 \le s_{f0} < H < s_{r0}$
As shown in Figure 5.3(b), the position X where $s_0(X)$ becomes larger than H for a positive angle of inclination of vehicle θ_{t0} can be calculated as:

$$X = \frac{H - s_{f0}}{\tan \theta_{t0}} = X_1$$

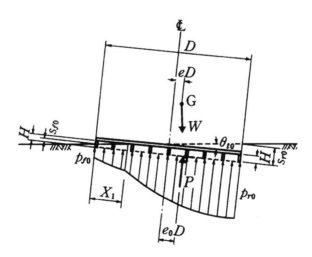

Figure 5.3(b). Distribution of contact pressure $P_i(X)$ and amount of sinkage $S_i(X)$ for $0 \le s_{f0} < H < s_{r0}$.

thence,

$$W \cos \theta_{t0} = 2B \int_0^{X_1} p_0(X) \, dX + 2B \int_{X_1}^{D} p_0(X) \, dX$$

$$= \frac{2BD}{s_{r0} - s_{f0}} \left[\frac{k_1}{n_1 + 1} (H^{n_1+1} - s_{f0}{}^{n_1+1}) \right.$$

$$\left. + (s_{r0} - H) \left\{ p_0 + \frac{k_2}{n_2 + 1} (s_{r0} - H)^{n_2} \right\} \right] \tag{5.24}$$

$$W \cos \theta_{t0} \left(\frac{1}{2} + e_0 \right) D = 2B \int_0^{X_1} p_0(X) X \, dX + 2B \int_{X_1}^{D} p_0(X) \, dX$$

$$= \frac{2k_1 BD^2}{(s_{r0} - s_{f0})^2} \left\{ \frac{H^{n_1+2} - s_{f0}{}^{n_1+2}}{n_1 + 2} - \frac{s_{f0}(H^{n_1+1} - s_{f0}{}^{n_1+1})}{n_1 + 1} \right\}$$

$$+ \frac{2BD^2}{(s_{r0} - s_{f0})^2} \left\{ \frac{k_2}{n_2 + 2} (s_{r0} - H)^{n_2+2} \right.$$

$$+ \frac{k_2}{n_2 + 1} (H - s_{f0}) (s_{r0} - H)^{n_2+1}$$

$$\left. + \frac{1}{2} p_0 (s_{r0}{}^2 - H^2) - p_0 s_{f0} (s_{r0} - H) \right\} \tag{5.25}$$

From above equations, s_{f0} and s_{r0} can be calculated. Substituting these values into Eqs. (5.20) and (5.21), θ_{t0} and e_0 can be determined. For this case we get:

$$p_{f0} = k_1 (s_{f0})^{n_1} \tag{5.26}$$

$$p_{r0} = p_0 + k_2 (s_{r0} - H)^{n_2} \tag{5.27}$$

(3) *For the case where $s_{f0} > H > s_{r0} \geq 0$*
As shown in Figure 5.3(c), the position X where $s_0(X)$ becomes less than H for a negative angle of inclination of vehicle θ_{t0} can be calculated as:

$$X = D + \frac{H - s_{r0}}{\tan \theta_{t0}} = X_2$$

thence,

$$W \cos \theta_{t0} = 2B \int_0^{X_2} p_0(X) \, dX + 2B \int_{X_2}^{D} p_0(X) \, dX$$

$$= \frac{2k_1 BD}{s_{r0} - s_{f0}} \cdot \frac{s_{r0}{}^{n_1+1} - H^{n_1+1}}{n_1 + 1} - \frac{2BD}{s_{r0} - s_{f0}}$$

$$\times (s_{f0} - H) \left\{ p_0 + \frac{k_2}{n_2 + 1} (s_{f0} - H)^{n_2} \right\} \tag{5.28}$$

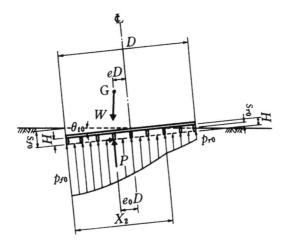

Figure 5.3(c). Distribution of contact pressure $P_i(X)$ and amount of sinkage $S_i(X)$ for $s_{f0} > H > s_{r0} \geq 0$.

$$W \cos \theta_{t0} \cdot \left(\frac{1}{2} + e_0\right) D = 2B \int_0^{X_2} p_0(X)\, dX + 2B \int_{X_2}^{D} p_0(X)X\, dX$$

$$= \frac{2BD^2}{(s_{r0} - s_{f0})^2} \left\{ -\frac{k_2}{n_2 + 2}(s_{f0} - H)^{n_2 + 2} - \frac{k_2 H}{n_2 + 1}(s_{f0} - H)^{n_2 + 1} \right.$$

$$+ \frac{k_2 s_{f0}}{n_2 + 1}(s_{f0} - H)^{n_2 + 1} + \frac{1}{2}p_0(H^2 - s_{f0}^2) - p_0 s_{f0}(H - s_{f0}) \bigg\}$$

$$+ \frac{2k_1 BD^2}{(s_{r0} - s_{f0})^2} \left\{ \frac{1}{n_1 + 2}(s_{r0}^{n_1 + 2} - H^{n_1 + 2}) \right.$$

$$- \frac{s_{f0}}{n_1 + 1}(s_{r0}^{n_1 + 1} - H^{n_1 + 1}) \bigg\} \tag{5.29}$$

From above equations, s_{f0} and s_{r0} can be calculated. Substituting these values into Eqs. (5.20) and (5.21), θ_{t0} and e_0 can be determined. The result in this case is

$$p_{f0} = p_0 + k_2(s_{f0} - H)^{n_2} \tag{5.30}$$

$$p_{r0} = k_1(s_{r0})^{n_1} \tag{5.31}$$

(4) *For the case where $s_{f0} < 0 < H < s_{r0}$*
As shown in Figure 5.3(d), the position X where $s_0(X)$ becomes larger than H for a positive angle of inclination of the vehicle θ_{t0} can be calculated as follows:

$$X = D - L + \frac{H}{\tan \theta_{t0}} = X_3$$

where L is given as follows:

$$L = \frac{s_{r0}}{s_{r0} - s_{f0}} D$$

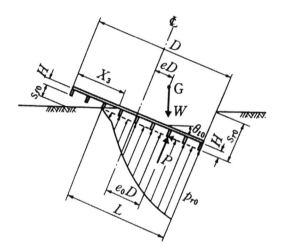

Figure 5.3(d). Distribution of contact pressure $P_i(X)$ and amount of sinkage $S_i(X)$ for $s_{f0} < H < 0 < s_{r0}$.

thence,

$$W \cos \theta_{t0} = 2B \int_{D-L}^{X_3} p_0(X)\,dX + 2B \int_{X_3}^{D} p_0(X)\,dX$$

$$= \frac{2k_1 BDH^{n_1+1}}{(n_1+1)(s_{r0}-s_{f0})} + \frac{2BD(s_{r0}-H)}{s_{r0}-s_{f0}} \times \left\{ p_0 + \frac{k_2}{n_2+1}(s_{f0}-H)^{n_2} \right\} \quad (5.32)$$

$$W \cos \theta_{t0} \cdot \left(\frac{1}{2} + e_0 \right) D = \frac{2k_1 BD^2}{(s_{r0}-s_{f0})^2} H^{n_1+1} \left(\frac{H}{n_1+2} - \frac{s_{f0}}{n_1+1} \right)$$

$$+ \frac{2BD^2}{(s_{r0}-s_{f0})^2} \left\{ \frac{1}{2} p_0(s_{r0}{}^2 - H^2) + k_2 (s_{r0}-H)^{n_2+1} \right.$$

$$\left. \times \left(\frac{s_{r0}-H}{n_2+2} + \frac{H-s_{f0}}{n_2+1} \right) - p_0 s_{f0}(s_{r0}-H) \right\} \quad (5.33)$$

From above equations, s_{f0} and s_{r0} can be calculated. Substituting these values into Eqs. (5.20) and (5.21), θ_{t0} and e_0 can be determined. In this case,

$$p_{f0} = 0 \quad (5.34)$$

$$p_{r0} = p_0 + k_2(s_{r0}-H)^{n_2} \quad (5.35)$$

(5) *For the case where $s_{f0} > H > 0 > s_{r0}$*
For the case shown in Figure 5.3(e), the position X where $s_0(X)$ is less than H for a negative angle of inclination of the vehicle θ_{t0} can be calculated as:

$$X = LL \left(1 - \frac{H}{s_{f0}} \right) = \frac{H - s_{f0}}{\tan \theta_{t0}} = X_4$$

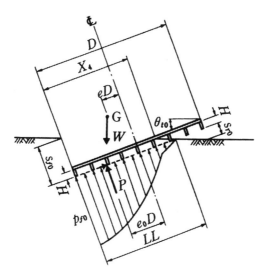

Figure 5.3(e). Distribution of contact pressure $P_i(X)$ and amount of sinkage $S_i(X)$ for $s_{f0} > H > 0 > s_{r0}$.

where LL is given as follows:

$$LL = \frac{s_{f0}}{s_{f0} - s_{r0}} D$$

thence,

$$
W \cos \theta_{t0} = 2B \int_0^{X_4} p_0(X)\, dX + 2B \int_{X_4}^{LL} p_0(X)\, dX
$$

$$
= \frac{2BD}{s_{f0} - s_{r0}} \left\{ p_0(s_{f0} - H) + \frac{k_1}{n_1 + 1} H^{n_1+1} + \frac{k_2}{n_2 + 1}(s_{f0} - H)^{n_2+1} \right\} \quad (5.36)
$$

$$
W \cos \theta_{t0} \cdot \left(\frac{1}{2} + e_0 \right) D = 2B \int_0^{X_4} p_0(X)\, dX + 2B \int_{X_4}^{LL} p_0(X)\, dX
$$

$$
= \frac{2BD^2}{(s_{f0} - s_{r0})^2} \left[p_0 s_{f0}(s_{f0} - H) + \frac{1}{2} p_0 (H^2 + s_{f0}^2) \right.
$$

$$
+ \frac{k_2 s_{f0}}{n_2 + 1}(s_{f0} - H)^{n_2+1} - k_2 \left\{ \frac{(s_{f0} - H)^{n_2+1}}{n_2 + 1} + \frac{H(s_{f0} - H)^{n_2+1}}{n_2 + 1} \right\}
$$

$$
\left. + k_1 \left(\frac{s_{f0}}{n_1 + 1} H^{n_1+1} - \frac{H^{n_1+2}}{n_1 + 2} \right) \right] \quad (5.37)
$$

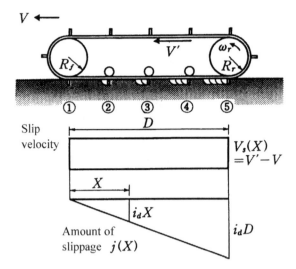

Figure 5.4. Distribution of slip velocity $V_s(X)$ and amount of slippage $j(X)$.

From above equations, the values of s_{f0} and s_{r0} can be determined. Substituting these values into Eqs. (5.20) and (5.21), θ_{t0} and e_0 can be calculated. For this case the results are:

$$p_{f0} = p_0 + k_2(s_{f0} - H)^{n_2} \tag{5.38}$$

$$p_{r0} = 0 \tag{5.39}$$

5.2   DRIVING STATE ANALYSIS

5.2.1   Amount of vehicle slippage

When a rigid tracked vehicle is running under traction during driving action on a hard terrain at a constant vehicle speed V and with a circumferential speed of track belt V' ($V \leq V'$), the slip velocity V_s of the track plate is constant for the whole range of the contact surface of the rigid track belt, as shown in Figure 5.4. However, it is evident that the amount of slippage j at each of the grouser positions must have a different value.

For a radius of front-idler R_f and of rear sprocket R_r, and given an angular velocity of the driving rear sprocket ω_r, a slip ratio i_d for the vehicle can defined as follows:

$$i_d = 1 - \frac{V}{V'} = \frac{R_r\omega_r - V}{R_r\omega_r} \tag{5.40}$$

The slip velocity $V_s(X)$ of the track plate at a distance X from the contact point B between the front-idler and the main part of the rigid track belt has a positive value, and can be calculated as follows:

$$V_s(X) = V' - V = R_r\omega_r - V \tag{5.41}$$

The leading grouser ① is assumed to move on the terrain only the amount of slippage j_0 during an elemental time interval Δt until the following grouser comes in contact with terrain subsequent to grouser ① contacting with terrain, as shown in the figure. During

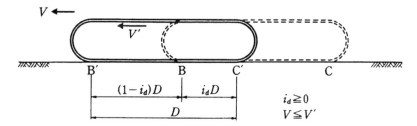

Figure 5.5. Amount of slippage of vehicle during driving action.

this time, the other grousers ②, ③, ④ and ⑤ connected to each others will move the same amount of slippage j_0 at the same time.

Therefore, the amount of slippage j of the nth grouser can be computed as the product of the proceeding time interval $n\Delta t$ after the beginning of contact and the slip velocity $j_0/\Delta t$ as follows:

$$j = \frac{j_0}{\Delta t} \cdot n\Delta t = nj_0 \tag{5.42}$$

Likewise, the amount of slippage $j(X)$ of the grouser at a distance X can be calculated by integrating the slip velocity V_s from zero to the required time $t = X/R_r\omega_r$ for the movement of the grouser from the point B to the point X as follows [5]:

$$j(X) = \int_0^t (R_r\omega_r - V)\,dt = \frac{R_r\omega_r - V}{R_r\omega_r} \cdot X = i_d X \tag{5.43}$$

Further, as the track belt moves a distance D around the vehicle itself during driving action, as shown in Figure 5.5, the amount of slippage at the rear end of the track belt $\overline{BC'}$ equals $i_d D$ while the movement of the vehicle $\overline{BB'}$ is $(1-i_d)D$, and the required time taken equals D/V'.

Thence, the required time t_d to the point when the vehicle has completely passed by the point B can be calculated as:

$$t_d = \frac{D}{V'} \cdot \frac{D}{(1 - i_d)D} = \frac{1}{1 - i_d} \cdot \frac{D}{V'} \tag{5.44}$$

The amount of slippage j_s of the point B on the terrain associated with the passage of the vehicle can be calculated as the product of the slip velocity V_s and the proceeding-time of the vehicle t_d as follows:

$$j_s = (V' - V)t_d = i_d V' \cdot \frac{1}{1 - i_d} \cdot \frac{D}{V'} = \frac{i_d D}{1 - i_d} \tag{5.45}$$

5.2.2 Force balance analysis

Figure 5.6 shows the composite of forces that act collectively on a rigid tracked vehicle when the vehicle is running under traction during driving action at a slip ratio i_d on a horizontal soft terrain. W is the vehicle weight, D is the contact length of the vehicle, R_f and R_r are the radius of the front-idler and rear sprockets respectively. H is defined as the grouser height, G_p is the grouser pitch, e is the eccentricity of the center of gravity of the vehicle, h_g is the height of the center of gravity G of the vehicle measured from the bottom

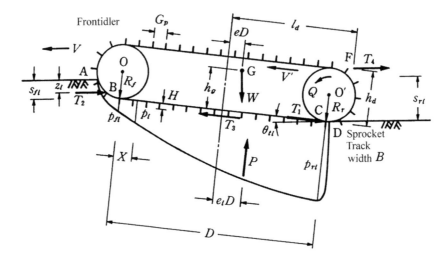

Figure 5.6. Forces acting on a rigid tracked vehicle during driving action.

of the track belt, l_d is the distance from the center line of the vehicle to the point F of application of the effective tractive effort, h_d is the height of the point F measured from the bottom of the track belt. V is the vehicle speed, V' is the circumferential speed of the track belt, and Q is the driving torque acting on the rear sprocket. The parameter s_{fi} is the amount of sinkage of the front-idler measured vertically at the contact point B between the front-idler and the main part of the track belt. Likewise, s_{ri} is the amount of sinkage of rear sprocket measured vertically at the contact point C between the rear sprocket and the main part of the track belt. The parameters p_{fi} and p_{ri} are the normal contact pressures acting at the points B and C of the track belt respectively.

The factor T_1 is the driving force acting on the circumferential lower part of the rear sprocket and $T_1 = Q/R_r$. T_2 is the compaction resistance which acts horizontally at a depth z_i on the front part of the track belt. T_3 is the thrust developed along the rigid track belt through each grouser as the sum of the shear resistances acting on the slip lines connecting each grouser tip. T_4 is the effective driving force which acts horizontally through the point F. P is the ground reaction acting normally to the contact part of the track belt. The eccentricity of P is e_i. The angle θ_{ti} is the angle of inclination of the vehicle. This angle can be calculated from the amounts of sinkage s_{fi} and s_{ri}.

The static amount of sinkage $s_0(X)$, taken as being normal to the main part of the rigid track belt at a distance X from the point B, can be expressed as a linear function of X using the static amount of sinkage s_{f0i} and s_{r0i} as follows:

$$s_{0i}(X) = s_{f0i} + (s_{r0i} - s_{f0i})\frac{X}{D} \qquad (5.46)$$

Thence, the distribution of the contact pressure $p_i(X)$ should be curved and it should take the following form:

For $0 \leq s_0(X) \leq H$

$$p_i(X) = k_1\{s_{0i}(X)\}^{n_1} \qquad (5.47)$$

For $s_0(X) > H$

$$p_i(X) = k_1 H^{n_1} + k_2\{s_{0i}(X) - H\}^{n_2} \tag{5.48}$$

The balance of forces acting on the vehicle in the horizontal and vertical directions and the moment balance around the rear axle can be expressed by the following equations:

$$T_2 + T_4 = T_3 \cos\theta_{ti} - P \sin\theta_{ti} \tag{5.49}$$

$$W = T_3 \sin\theta_{ti} + P \cos\theta_{ti} \tag{5.50}$$

$$Q = T_1 R_r = T_3 R_r \tag{5.51}$$

$$\theta_{ti} = \sin^{-1}\left(\frac{s_{ri} - s_{fi}}{D}\right) \tag{5.52}$$

Then, the effective driving force T_4 and the ground reaction P can be calculated as:

$$T_4 = \frac{T_3}{\cos\theta_{ti}} - W \tan\theta_{ti} - T_2 \tag{5.53}$$

$$P = \frac{W}{\cos\theta_{ti}} - T_3 \tan\theta_{ti} \tag{5.54}$$

For $i_d = 0$, the above equation will be satisfied if $T_2 = T_4 = 0$ and $T_3 = W \sin\theta_{t0}, \theta_{ti} = \theta_{t0}$ and $P = W \cos\theta_{t0}$.

Next, for the pure rolling state of the vehicle, it is proposed that $T_1 = T_3 = W \sin\theta_{ti}$, $T_4 = -T_2$ and $P = W \cos\theta_{ti}$, and that the depth z_i where the compaction resistance T_2 acts can be calculated using the rut depth of the vehicle e.g. the amount of sinkage of rear sprocket s_{ri} and the angle of inclination of vehicle θ_{ti}, as follows:

$$Z_i = S_{ri} - \left[T_2\left\{R_r + (h_d - R_r)\cos\theta_{ti} - \left(l_d - \frac{D}{2}\right) \times \sin\theta_{ti}\right\}\right.$$
$$\left. -W\left\{h_g \sin\theta_{ti} - D\left(\frac{1}{2} - e\right)\cos\theta_{ti}\right\}\right] / \left(T_2 + \frac{W\cos\theta_{ti}}{\tan\theta_{ti}}\right) \tag{5.55}$$

We assume here that the moment around the axle of the rear sprocket due to T_2, T_3, P and W are zero or are negligibly small.

Additionally, the moment balance equation operating as a result of a composite of forces acting on the rigid tracked vehicle during driving action with slip ratio i_d can be expressed as:

$$-Q + T_3 R_r - T_2(R_r - s_{ri} + z_i) + T_4\left\{(h_d - R_r)\cos\theta_{ti} - \left(l_d - \frac{D}{2}\right)\sin\theta_{ti}\right\}$$
$$+W\left\{(h_g - R_r)\sin\theta_{ti} - D\left(\frac{1}{2} - e\right)\cos\theta_{ti}\right\} + PD\left(\frac{1}{2} - e_i\right) = 0 \tag{5.56}$$

Substituting the previous Eq. (5.51) into the above equation, the expression can be solved for eccentricity e_i as follows:

$$e_i = \frac{1}{2} + \frac{1}{PD}\left[-T_2(R_r - s_{ri} + z_i) + T_4\left\{(h_d - R_r)\cos\theta_{ti} - \left(l_d - \frac{D}{2}\right)\sin\theta_{ti}\right\}\right.$$
$$\left. + W\left\{(h_g - R_r)\sin\theta_{ti} - D\left(\frac{1}{2} - e\right)\cos\theta_{ti}\right\}\right] \tag{5.57}$$

For $i_d = 0$, $P = W\cos\theta_{t0}$ and $T_2 = T_4 = 0$, when a rigid tracked vehicle is stopped under conditions of $T_3 = W\sin\theta_{t0}$ and $Q = WR_r\sin\theta_{t0}$. In this case, the eccentricity e_0 of the ground reaction P can be calculated as:

$$e_0 = e + \frac{h_g - R_r}{D}\tan\theta_{t0} \tag{5.58}$$

This notion becomes important when the vehicle is climbing up a slope.

It is noted here that the value of e_0 given in Eq. (5.21) is different from the above mentioned value because the rigid track belt at rest was calculated for $Q = 0$ assuming that the track belt is a rotational rigid body supported by the bearing capacity of terrain at the front and rear contact part or by the braking action.

5.2.3   Thrust analysis

As shown in Figure 5.7, the thrust T_3 developed on the contact part of a track belt [6] is comprised of the sum T_{mb} of the shear resistances $\tau_m(X)$ acting on the base area of grousers of the main part of track belt, and the sum T_{ms} of the shear resistances $\tau_{ms}(X)$ acting on the side parts of the grousers. Additionally there are components representing the sum T_{fb}

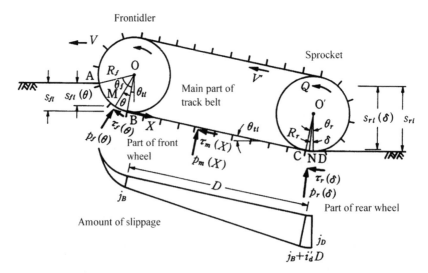

Figure 5.7. Stresses acting on main part of track belt, parts of front and rear wheel and distribution of amount of slippage (during driving action).

of the shear resistances $\tau_f(\theta)$ acting on the base area of grousers of the contact part of the front-idler, the sum T_{fs} of the shear resistances τ_{fs} (θ) acting on the side parts of the grousers, and the sum T_{rb} of the shear resistances $\tau_r(\delta)$ acting on the base area of grousers of the contact part of the rear sprocket. These forces are supplemented by the summation T_{rs} of shear resistances $\tau_{rs}(\delta)$ acting on the side parts of the grousers, which develop with increasing amounts of sinkage. Overall the total picture is as follows;

$$T_3 = T_{mb} + T_{ms} + T_{fb} + T_{fs} + T_{rb} + T_{rs} \qquad (5.59)$$

where X is the distance from the point B on the main part of the track belt, θ is the central angle $\angle\text{BOM}$ for an arbitrary contact point M on the front-idler, and δ is the central angle $\angle\text{DO}'\text{N}$ between the end point D and an arbitrary point N on the contact part of the rear sprocket.

(1) *Main part of track belt [7]*
The amount of static sinkage $s_{0i}(X)$ of the main part of track belt BC and the distribution of the contact pressure $p_i(X)$ can be calculated from Eqs. (5.46) and (5.47), (5.48) respectively. Since the main part of the track belt is inclined at a vehicle inclination angle of θ_{ti} during driving action, the slip ratio i'_d needs to be modified using i_d given in Eq. (5.40). This may be done as follows:

$$i'_d = 1 - \frac{V}{\cos\theta_{ti}} \cdot \frac{1}{V'} = 1 - \frac{1 - i_d}{\cos\theta_{ti}} \qquad (5.60)$$

Thence, the distribution of the amount of slippage $j(X)$ is given for an amount of slippage j_B at the point B as:

$$j(X) = j_B + i'_d X \qquad (5.61)$$

The distribution of shear resistance $\tau_m(X)$ acting on the base area of the grousers of the main part of track belt can be calculated from the previous Eq. (4.5) as follows:

$$\tau_m(X) = \{m_c + m_f \cdot p_i(X)\}\left[1 - \exp\{-a(j_B + i'_d X)\}\right] \qquad (5.62)$$

Then, the overall thrust T_{mb} acting on the base area of grousers of the main part of the track belt can be calculated as:

$$T_{mb} = 2B \int_0^D \tau_m(X)\,dX \qquad (5.63)$$

Further, the distribution of the shear resistance $\tau_{ms}(X)$ that acts on the four sides of the grousers of the main part of track belt can be calculated from the average normal stress F_x/DH acting on the side part of the track belt given in Eq. (5.14) as:

$$t_{ms}(X) = 4H\left\{m_c + m_f \cdot \frac{p_i(X)}{\pi}\cot^{-1}\left(\frac{H}{B}\right)\right\} \times \left[1 - \exp\{-a(j_B + i'_d X)\}\right] \qquad (5.64)$$

Then, the thrust T_{ms} acting on the side part of grousers of the main track belt can be determined as:

$$T_{ms} = \int_0^D t_{ms}(X)\,dX \qquad (5.65)$$

Since the above calculations correspond with $s_{f0} = s_{f0i}$, $s_{r0} = s_{r0i}$ in the cases of the Section 5.1.2 (1), (2), (3), $p_i(X)$ can be calculated for each amount of sinkage $s_{0i}(X)$.

Now, let us consider the situation where $s_{f0i} < 0 < H < s_{r0i}$ as in Section 5.1.2(4). Substituting $W \cos \theta_{t0} = P$, $p_0(X) = p_i(X)$ and $e_i = e_0$ into Eqs. (5.32), (5.33), the amount of sinkage $s_{f0} = s_{f0i}$ and $s_{r0} = s_{r0i}$ can be calculated. The real contact length L of the main part of the track belt is thence calculated as:

$$L = \frac{s_{r0i}}{s_{r0i} - s_{f0i}} D \tag{5.66}$$

For $0 \leq X \leq D - L$ the contact pressure $p_i(X)$ becomes zero and then the shear resistance $\tau_m(X)$ becomes zero. For $D - L < X \leq D$, $p_i(X)$ can be calculated from Eqs. (5.47), (5.48) by use of the amount of sinkage $s_{0i}(X)$ given in Eq. (5.46) and then $\tau_m(X)$ can be determined from the following expression.

$$\tau_m(X) = \left\{ m_c + m_f \cdot p_i(X) \right\} \left[1 - \exp\{ -a i'_d (X - D + L) \} \right] \tag{5.67}$$

Then, the thrust T_{mb} acting on the base area of the grousers of the main contact part of the track belt can be calculated as follows:

$$T_{mb} = 2B \int_{D-L}^{D} \tau_m(X) \, dX \tag{5.68}$$

Following this, the shear resistance $\tau_{ms}(X)$ acting on the four sides of the grousers of the main contact part of the track belt may be calculated as:

$$t_{ms}(X) = 4H \left\{ m_c + m_f \cdot \frac{p_i(X)}{\pi} \cot^{-1} \left(\frac{H}{B} \right) \right\} \times \left[1 - \exp\{ -a i'_d (X - D + L) \} \right] \tag{5.69}$$

Next, the thrust T_{ms} acting on the side of grousers of the main contact part of the track belt can be calculated as follows:

$$T_{ms} = \int_{D-L}^{D} t_{ms}(X) \, dX \tag{5.70}$$

Further, let us consider the case where $s_{f0i} > H > s_{r0i}$ as in Section 5.1.2(5). Calculating the amount of sinkage s_{f0i} and s_{r0i} from Eqs. (5.36), (5.37) in the same manner as previously, the real contact length LL of the main part of the track belt can be calculated as follows:

$$LL = \frac{s_{f0i}}{s_{f0i} - s_{r0i}} D \tag{5.71}$$

For $LL < X \leq D$, the contact pressure $p_i(X)$ becomes zero and then the shear resistance $\tau_m(X)$ becomes zero. For $0 \leq X \leq LL$, $p_i(X)$ can be calculated from Eqs. (5.47), (5.48) by use of the amount of sinkage $s_{0i}(X)$ given in Eq. (5.46) and then $\tau_m(X)$ can be given as follows:

$$\tau_m(X) = \left\{ m_c + m_f \cdot p_i (X) \right\} \left[1 - \exp\{ -a(j_B + i'_d X) \} \right] \tag{5.72}$$

Then, the thrust T_{mb} acting on the base area of grousers of the main contact part of the track belt can be calculated as follows:

$$T_{mb} = 2B \int_0^{LL} \tau_m(X) \, dX \tag{5.73}$$

Moreover, the shear resistance $t_{ms}(X)$ acting on the four sides of grousers of the main contact part of the track belt is given as:

$$t_{ms}(X) = 4H \left\{ m_c + m_f \cdot \frac{p_i(X)}{\pi} \cot^{-1}\left(\frac{H}{B}\right) \right\} [1 - \exp\{-a(j_B + i'_d X)\}] \tag{5.74}$$

thence, the thrust T_{ms} acting on the side parts of the grousers of the main contact part of the track belt can be calculated as:

$$T_{ms} = \int_0^{LL} t_{ms}(X) \, dX \tag{5.75}$$

(2) *Contact part of front-idler*

As the front-idler comes into contact with the terrain at point A (as shown in Figure 5.7), the entry angle $\theta_f = \angle AOB$ can be calculated as follows:

$$\theta_f = \cos^{-1}\left(\cos\theta_{ti} - \frac{s_{fi}}{R_f + H} \right) - \theta_{ti} \tag{5.76}$$

The amount of slippage $j_f(\theta)$ at the contact part of the front-idler can be calculated through the use of the entry angle θ_{f0} and the angle of inclination of vehicle θ_{t0i} (which in turn is calculated from the static amounts of sinkage s_{f0i} and s_{r0i}). Using these ideas we get:

$$j_f(\theta) = (R_f + H) \int_\theta^{\theta_{f0}} \left\{ 1 - (1 - i_d) \cos(\theta + \theta_{t0i}) \right\} d\theta$$
$$= (R_f + H) \left[(\theta_{f0} - \theta) - (1 - i_d) \{ \sin(\theta_{f0} + \theta_{t0i}) - \sin(\theta + \theta_{t0i}) \} \right] \tag{5.77}$$

$$\theta_{f0} = \cos^{-1}\left(\cos\theta_{t0i} - \frac{s_{f0i}\cos\theta_{t0i}}{R_f + H} \right) - \theta_{t0i} \tag{5.78}$$

$$\theta_{t0i} = \tan^{-1}\left(\frac{s_{r0i} - s_{f0i}}{D} \right) \tag{5.79}$$

In addition, the amount of slippage j_B at the point B is given as $j_f(0)$ from the above Eq. (5.77).

The amount of sinkage $s_{f0i}(\theta)$ at the contact part of the front-idler is given as follows:

$$s_{f0i}(\theta) = \frac{(R_f + H)\{\cos(\theta + \theta_{t0i}) - \cos(\theta_{f0}\theta_{t0i})\}}{\cos(\theta + \theta_{t0i})} \tag{5.80}$$

then the contact pressure $p_{fi}(\theta)$ can be calculated as in the following equations:

For $0 \leq s_{f0i}(\theta) \leq H$

$$p_{fi}(\theta) = k_1 \{s_{f0i}(\theta)\}^{n_1} \tag{5.81}$$

For $s_{f0i}(\theta) > H$

$$p_{fi}(\theta) = k_1 H^{n_1} + k_2 \{s_{f0i}(\theta) - H\}^{n_2} \qquad (5.82)$$

Then, the shear resistance $\tau_f(\theta)$ acting on the contact part of the front-idler can be calculated as follows:

$$\tau_{fi}(\theta) = \{m_c + m_f \cdot p_{fi}(\theta)\} [1 - \exp\{-aj_f(\theta)\}] \qquad (5.83)$$

Thence, the thrust T_{fb} acting on the base area of grousers of the front-idler in the direction of the main part of the track belt BC can be determined as follows:

$$T_{fb} = 2B(R_{fg} + H) \int_0^{\theta f 0} \{\tau_f(\theta)\cos\theta - p_{fi}(\theta)\sin\theta\} \, d\theta \qquad (5.84)$$

Further, the shear resistance $t_{fs}(\theta)$ acting on the four sides of the grousers on the front-idler is given as:

$$t_{fs}(\theta) = 4H \left\{ m_c + m_f \cdot \frac{p_{fi}(\theta)}{\pi} \cot^{-1}\left(\frac{H}{B}\right) \right\} [1 - \exp\{-aj_f(\theta)\}] \qquad (5.85)$$

then the thrust T_{fs} acting on the side part of the front-idler can be calculated:

$$T_{fs} = (R_f + H) \int_0^{\theta f 0} t_{fs}(\theta)\cos\theta d\theta \qquad (5.86)$$

The above calculations correspond to the cases of Section 5.1.2(1), (2), (3) and (5). For the case of Section 5.1.2 (4), both the values of T_{fb} and T_{fs} become zero.

(3)   Contact part of rear sprocket

As the rear sprocket separates finally from the terrain at a point D as shown in Figure 5.7, the exit angle $\theta_r = \angle CO'D$ is almost equal to the angle of inclination of the vehicle θ_{ti} providing that we assume that the expansive amount of deformation of the terrain due to unloading is negligibly small.

The amount of slippage $j_r(\delta)$ at the contact part of the rear sprocket can be calculated through use of the exit angle $\theta_{r0} = \theta_{t0i}$ and the angle of inclination of the vehicle θ_{t0i} which in turn is calculated from the static amount of sinkage of the front and rear wheel s_{f0i} and s_{r0i}. The calculation equation is as follows:

$$j_r(\delta) = (R_r + H)\{(\theta_{t0i} - \delta) - (1 - i_d)(\sin\theta_{t0i} - \sin\delta)\} + i'_d D + j_B \qquad (5.87)$$

Additionally, the amount of slippage j_D at the point D is given as $j_r(0)$ from the above equation.

The amount of sinkage $s_{r0i}(\delta)$ at the contact part of the rear sprocket is given as follows:

$$s_{r0i}(\delta) = [(R_r + H)\{\cos\theta_{t0i} - \cos(\theta_{f0} + \theta_{t0i})\} + D\sin\theta_{t0i}$$
$$+ (R_r + H)(\cos\delta - \cos\theta_{t0i})] / \cos\delta \qquad (5.88)$$

Thence the contact pressure $p_{ri}(\delta)$ can be calculated as follows:

For $0 \leq s_{r0i}(\delta) \leq H$

$$p_{ri}(\delta) = k_1 \{s_{r0i}(\delta)\}^{n_1} \tag{5.89}$$

For $s_{r0i}(\delta) > H$

$$p_{ri}(\delta) = k_1 H^{n_1} + k_2 \{s_{r0i}(\delta) - H\}^{n_2} \tag{5.90}$$

Following this, the shear resistance $\tau_r(\delta)$ acting on the contact part of the rear sprocket can be determined by the following expression.

$$\tau_r(\delta) = \{m_c + m_f \cdot p_{ri}(\delta)\}[1 - \exp\{-aj_r(\delta)\}] \tag{5.91}$$

Thence, the thrust T_{rb} acting on the base area of the grousers of the rear sprocket in the direction of the main part of the track belt BC can be calculated as:

$$T_{rb} = 2B(R_r + H) \int_0^{\theta r0} \{\tau_r(\delta)\cos(\theta_{t0i} - \delta) + p_{ri}(\delta)\sin(\theta_{t0i} - \delta)\} \, d\delta \tag{5.92}$$

Additionally, the shear resistance $t_{rs}(\delta)$ acting on the four sides of the grousers of the rear sprocket is given as:

$$t_{rs}(\delta) = 4H \left\{ m_c + m_f \cdot \frac{p_{ri}(\delta)}{\pi} \cot^{-1}\left(\frac{H}{B}\right) \right\} [1 - \exp\{-aj_r(\delta)\}] \tag{5.93}$$

Then, the thrust T_{rs} acting on the side parts of the grousers of the rear sprocket can be calculated as:

$$T_{rs} = (R_r + H) \int_0^{\theta r0} t_{rs}(\delta)\cos(\theta_{t0i} - \delta) \, d\delta \tag{5.94}$$

The above calculations correspond to the cases of Sections 5.1.2 (1), (2), (3) and (5). In the case of Section 5.1.2 (4), the values of T_{rb} and T_{rs} can be determined by substituting the following expressions into Eqs. (5.91), (5.92) and (5.93), (5.94), respectively.

$$j_r(\delta) = (R_r + H)\{(\theta_{t0i} - \delta) - (1 - i_d)(\sin\theta_{t0i} - \sin\delta)\} + i'_d L \tag{5.95}$$

$$s_{r0i}(\delta) = \frac{L \sin\theta_{t0i} + (R_r + H)(\cos\delta - \cos\theta_{t0i})}{\cos\delta} \tag{5.96}$$

For the above calculations of the thrust of a rigid tracked vehicle running on a flat terrain, the constant amount of slippage j_w which occurs on the whole range of contact part of the track belt due to the force $W \sin\theta_{ti}$ is assumed to have a negligibly small value.

5.2.4   Compaction resistance

When a rigid tracked vehicle is running on a horizontal soft terrain, it makes a rut of depth s_{ri} which corresponds to the amount of sinkage of the rear sprocket. To calculate the depth of this rut, the amount of sinkage of the rigid tracked vehicle has to be considered during the driving state.

The total amount of sinkage s_{fi} and s_{ri} of a rigid track belt can be calculated from the static amounts of sinkage s_{f0i} and s_{r0i} and the amounts of slip sinkage s_{fs} and s_{rs} of front and rear wheel (which are normal to the main part of the track belt) as follows:

$$s_{fi} = (s_{f0i} + s_{fs})\cos\theta_{ti} \tag{5.97}$$

$$s_{ri} = (s_{r0i} + s_{rs})\cos\theta_{ti} \tag{5.98}$$

where s_{f0i} and s_{r0i} equal $s_{0i}(0)$ and $s_{0i}(D)$ – respectively calculated from Eq. (5.46) – and θ_{ti} is given by Eq. (5.52).

The amount of slip sinkage s_{fs} at the point B on the front-idler can be calculated by substituting the contact pressure $p_{fi}(\theta_m)$ and the amount of slippage of the soil j_{fs} into the previous Eq. (4.8). This yields the expression:

$$s_{fs} = c_0 \sum_{m=1}^{M} \{p_{fi}(\theta_m)\cos\theta_m\}^{c_1} \left\{ \left(\frac{m}{M} j_{fs}\right)^{c_2} - \left(\frac{m-1}{M} j_{fs}\right)^{c_2} \right\} \tag{5.99}$$

where

$$\theta_m = \theta_{f0}\left(1 - \frac{m}{M}\right)$$

The parameter θ_{f0} can be calculated from Eq. (5.78) and $p_{fi}(\theta_m)$ can be calculated from Eqs. (5.81) and (5.82). The amount of slippage of soil j_{fs} of the contact part of the front-idler can be determined as:

$$j_{fs} = (R_r + H)\sin\theta_f \cdot \frac{i_d}{1 - i_d} \tag{5.100}$$

Following this, the amount of slip sinkage s_{rs} at the point C on the rear sprocket can be calculated as follows [8] by substituting the contact pressure $p_i(X)$ and the amount of slippage of soil j_s into the previous Eq. (4.8). Doing this, we get:

$$s_{rs} = s_{fs} + c_0 \sum_{n=1}^{N} \left\{ p_i \left(\frac{nD}{N}\right) \right\}^{c_1} \left[\left(\frac{nj_s}{N}\right)^{c_2} - \left\{ \frac{(n-1)j_s}{N} \right\}^{c_2} \right] \tag{5.101}$$

where

$$j_s = \frac{i'_d D}{1 - i'_d}$$

Furthermore, the total amount of sinkage $s_i(X)$ of the rigid track belt needs to satisfy the following equation.

$$s_i(X) = s_{fi} + (s_{ri} - s_{fi})\frac{X}{D} \tag{5.102}$$

The compaction resistance T_2 can be calculated from the total amount of sinkage s_{ri} of the rear sprocket obtained from above equations. That is, as shown in Figure 5.8, when a rigid tracked vehicle of track width B and contact length D moves through a distance L_x, the work $T_2 L_x$ to make the rut of $\overline{ABDD'A'}$ can be assumed to be equal the work necessary to

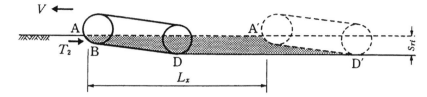

Figure 5.8. Work due to compaction resistance T_2.

compact the terrain of the area of $2BL_X$ corresponding to the contact part of track belt to the depth s_{ri} as follows:

$$T_2 \cdot L_x = 2B \cdot L_x \int_0^{s_{ri}} p \, ds_0$$

Now, using the previous Eqs. (4.1) and (4.2), the compaction resistance T_2 can be calculated as:

$$T_2 = 2B \left[\int_0^H k_1 s_0^{n_1} \, ds_0 + \int_H^{s_{ri}} \left\{ k_1 H^{n_1} + k_2 (s_0 - H)^{n_2} \right\} ds_0 \right]$$

$$= \frac{2k_1 B}{n_1 + 1} H^{n_1 + 1} + 2k_1 B H^{n_1}(s_{ri} - H) + \frac{2k_2 B}{n_2 + 1}(s_{ri} - H)^{n_2 + 1} \tag{5.103}$$

The above calculations can be applied in the case of $0 \leq s_{f0i} \leq s_{r0i}$.

In the situation where $s_{f0i} > s_{r0i} > 0$, the total amount of sinkage of the rear sprocket following the passage of the front-idler should, of necessity, be larger than the total amount of sinkage of the front-idler. Therefore, Eq. (5.98) can be rewritten as follows:

$$s_{ri} = (s_{f0i} + s_{r0i}) \cos \theta_{ti} = s_{fi} + (s_{rs} - s_{fs}) \cos \theta_{ti} \tag{5.104}$$

Moreover, for the situation where $s_{f0i} < 0 < H < s_{r0i}, s_{fs} = 0$ and s_{fs} can be calculated, by use of the factor L given in Eq. (5.66), as follows:

$$s_{rs} = c_0 \sum_{n=1}^{N} \left\{ p_i \left(\frac{nL}{N} \right) \right\}^{c_1} \left[\left(\frac{nj_s}{N} \right)^{c_2} - \left\{ \frac{(n-1)j_s}{N} \right\}^{c_2} \right] \tag{5.105}$$

where

$$j_s = \frac{i_d' L}{1 - i_d'}$$

Thence, the total amount of sinkage s_{fi} and s_{ri} can be calculated by substituting the values of s_{fs} and s_{rs} obtained from the above equation into Eqs. (5.97), (5.98), respectively.

Next, for the situation where $s_{f0i} > H > 0 > s_{r0i}, s_{fs}$ and s_{fi} can be calculated respectively from Eqs. (5.99), (5.97). The parameter s_{rs} can be calculated, by use of the parameter LL given in Eq. (5.71), as follows:

$$s_{rs} = s_{fs} + c_0 \sum_{n=1}^{N} \left\{ p_i \left(\frac{nLL}{N} \right) \right\}^{c_1} \left[\left(\frac{nj_s}{N} \right)^{c_1} - \left\{ \frac{(n-1)j_s}{N} \right\}^{c_2} \right] \tag{5.106}$$

where

$$j_s = \frac{i'_d LL}{1 - i'_d}$$

After this, the total amount of sinkage of rear sprocket s_{ri} following the passage of the front-idler can be calculated by use of Eq. (5.104).

For each of the situations mentioned above, the compaction resistance T_2 can be calculated by substituting the obtained total amount of sinkage of the rear sprocket s_{ri} into Eq. (5.103).

5.2.5   Energy equilibrium equation

The effective input energy E_1 supplied by a driving torque applied to the rear sprocket must equals the total output energy i.e. the sum of the compaction energy E_2 required to make a rut under the track belt, the slippage energy E_3 that occurs during shear deformation of the soil under the track belt plus the effective driving force energy E_4 which is the work done by the drawbar pull of the rigid tracked vehicle. Thus the following energy equilibrium equation can be set up.

$$E_1 = E_2 + E_3 + E_4 \tag{5.107}$$

Next, let us consider the situation where a track belt moves a distance of contact length D of in relation to the rigid tracked vehicle. As the vehicle moves horizontally the distance of $(1 - i'_d)D \cos \theta_{ti}$ during this time, the various energy components can be calculated as follows:

$$E_1 = T_1 D = T_3 D \tag{5.108}$$

$$E_2 = T_2(1 - i'_d)D \cos \theta_{ti} = T_2(1 - i_d)D \tag{5.109}$$

$$E_3 = T_3 i'_d D + W(1 - i'_d)D \sin \theta_{ti} = T_3 \left(1 - \frac{1 - i_d}{\cos \theta_{ti}} \right) D + W(1 + i_d)D \tan \theta_{ti} \tag{5.110}$$

$$E_4 = T_4(1 - i'_d)D \cos \theta_{ti} = T_4(1 - i_d)D \tag{5.111}$$

Further, each energy component per unit of time can be calculated using the vehicle speed V and the circumferential speed of track belt V' as follows:

$$E_1 = T_1 V' = T_3 V' = T_3 \frac{V}{1 - i_d} \tag{5.112}$$

$$E_2 = T_2(1 - i_d)V' = T_2 V \tag{5.113}$$

$$E_3 = T_3 \left(1 - \frac{1 - i_d}{\cos \theta_{ti}} \right) V' + W(1 - i_d)V' \tan \theta_{ti}$$

$$= T_3 \left(\frac{1}{1 - i_d} - \frac{1}{\cos \theta_{ti}} \right) V + WV \tan \theta_{ti} \tag{5.114}$$

$$E_4 = T_4(1 - i_d)V' = T_4 V \tag{5.115}$$

5.2.6   Effective driving force

The effective driving force T_4 that a rigid tracked vehicle develops during driving action can be calculated by substituting the thrust T_3, the compaction resistance T_2 and the angle of inclination of the vehicle θ_{ti} into Eq. (5.53). In the relationship between the effective driving force T_4 and slip ratio i_d, an 'optimum slip ratio' i_{dopt} is defined as the slip ratio at which the effective driving force energy E_4 takes on a maximum value for a constant circumferential speed of track belt V'. Likewise an 'optimum effective driving force' T_{dopt} is defined as the effective driving force corresponding to the optimum slip ratio. Additionally, a 'tractive power efficiency' factor E_d can be defined as:

$$E_d = \frac{E_4}{E_1} = \frac{VT_4}{V'T_1} = (1 - i_d)\frac{T_4}{T_1} \tag{5.116}$$

Figure 5.9 presents a flow chart that may be used to calculate the tractive performance of a rigid tracked vehicle running on a soft terrain during driving action. Initially, the geometrical dimensions of the vehicle W, B, D, R_f, R_r, H, e, h_g, l_d, h_d and V' are required as input data. Given these, the terrain-track system constants k_1, n_1, k_2, n_2, m_c, m_f, a, c_0, c_1, and c_2 can be read. After this, the contact pressure distributions p_{f0}, p_{r0} and $p_0(X)$ of the rigid tracked vehicle at rest can be calculated for the static amount of sinkage distributions s_{f0}, s_{r0} and $s_0(X)$ and the angle of inclination of the vehicle θ_{t0} for each of the cases (1) $\sim$ (5) as mentioned in Section 5.1.2. This factor should be calculated iteratively until the magnitude of the sinkage $s_0(X)$ is strictly determined. Next, the slip ratio i_d during driving action can be taken from $i_d = 1\%$ to 99% successively.

The calculation flow system can be divided into the following three streams or strands according to the static amount of sinkage of the front-idler s_{f0i} and the rear sprocket s_{r0i}.

For the situation where $s_{f0i} \geq H$, $s_{r0i} \geq H$; $0 \leq s_{f0i} < H < s_{r0i}$; $s_{f0i} > H > s_{r0i} \leq 0$, $s_{0i}(X)$ and θ_{tio} can be calculated from Eqs. (5.46) and (5.47) respectively. Then p_{fi}, p_{ri} and $p_i(X)$ can be calculated using Eqs. (5.47) and (5.48). After this, the amounts of slip sinkage s_{fs} and s_{rs} can be calculated from Eqs. (5.99) and (5.101) respectively. Thence, the total amount of sinkage of the front-idler s_{fi} and the rear sprocket s_{ri} can be calculated from Eqs. (5.97) and (5.98) respectively in the case of $0 \leq s_{f0i} \leq s_{r0i}$, and from Eqs. (5.97) and (5.104) respectively in the case of $s_{f0i} > s_{r0i} > 0$. Following this, $s_0(X)$ and θ_{ti} can be determined by use of Eqs. (5.102) and (5.52), respectively.

In a next step, the compaction resistance T_2 can be calculated by substituting s_{ri} into Eq. (5.103). The thrust T_3 can then be calculated from Eq. (5.59) as the sum of the factor T_{mb} given in Eq. (5.63), T_{ms} given in Eq. (5.65), T_{fb} given in Eq. (5.84), T_{fs} given in Eq. (5.86), T_{rb} given in Eq. (5.92) and T_{rs} given in Eq. (5.94). Following this, the driving force T_1 given in Eq. (5.51), the effective driving force T_4 given in Eq. (5.53) and the ground reaction P given in Eq. (5.54) can be calculated by iteration until the thrust T_3 is uniquely defined.

It should be noted that in the process of calculation of the static amounts of sinkage of the front-idler and rear sprockets, the ground reaction P' acting on the main part of the rigid track belt should be calculated by subtracting the following ground reaction P_f acting on the contact part of the front-idler from the above mentioned ground reaction P.

$$P_f = 2B(R_f + H) \int_0^{\theta_{f0}} \left\{ \tau_f(\theta) \sin\theta + p_{fi}(\theta) \cos\theta \right\} d\theta \tag{5.117}$$

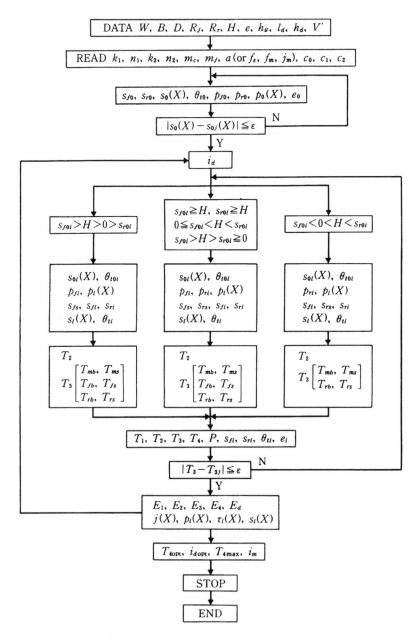

Figure 5.9. Flow chart to calculate the tractive performance of a rigid tracked vehicle running on soft terrain during driving action.

where $\tau_f(\theta)$ and $p_{fi}(\theta)$ may be calculated from Eqs. (5.83) and (5.81), (5.82) respectively and θ_{f0} can be calculated from Eq. (5.78).

In addition, in the cases (1), (2), (3) given in the Section 5.1.2, s_{f0i} and s_{r0i} may be calculated by substituting the value of $P' = P - P_f$ into $W \cos \theta_{t0}$.

For cases where $s_{f0i} < 0 < H < s_{r0i}$, $s_{0i}(X)$ and θ_{t0i} and p_{fi} ($=0$), p_{ri}, $p_i(X)$ can be calculated in a similar manner. The amount of slip sinkage of the rear sprocket s_{ri} can be calculated from Eq. (5.105), while that of the front-idler s_{fs} is zero. Thence, the total amount of sinkage of the front-idler s_{fi} and the rear sprocket s_{ri} can be calculated from Eqs. (5.97) and (5.98) respectively and then $s_i(X)$ and θ_{ti} can be determined.

Following this, the compaction resistance T_2 can be calculated from the obtained amount of sinkage s_{ri}. The thrust T_3 can be calculated as the sum of the elements T_{mb} given in Eq. (5.68), T_{ms} given in Eq. (5.70), T_{rb} given in Eq. (5.92) and T_{rs} given in Eq. (5.94). The driving force T_1, the effective driving force T_4 and the ground reaction P can be calculated iteratively until the thrust T_3 is uniquely defined.

For the case where $s_{f0i} > H > 0 > s_{r0i}$, $s_{0i}(X)$ and θ_{t0i} and p_{fi}, p_{ri} ($=0$), $p_i(X)$ can be calculated in the same way. The amount of slip sinkage of the front-idler s_{fs} can be calculated from Eq. (5.99) and the amount of slip sinkage of the rear sprocket s_{rs} can be calculated from Eq. (5.106). Thence, the total amount of sinkage of the front-idler s_{fi} and the rear sprocket s_{ri} can be calculated from Eqs. (5.97) and (5.104) respectively and then $s_i(X)$ and θ_{ti} can be determined.

Next, the compaction resistance T_2 can be calculated from the resultant amount of sinkage s_{ri}.

In the calculation of the thrust T_3, the value of the factors T_{mb} given in Eq. (5.68) and the T_{ms} given in Eq. (5.70) as the sum of the shear resistance of the soil acting on the base and areas of grousers of the main part of the track belt, and the factors T_{rb} given in Eq. (5.92) and T_{rs} given in Eq. (5.94) as the sum of soil acting on the base and side areas of grousers of the contact part of the rear sprocket are required to be modified. This modification derives from considering the fact that that the rear part of the track belt that follows the passage of the front-idler comes into contact with the terrain. Then the contact part which lies in the range of $LL < X < D$ develops some cohesive resistance – given as a function of m_c – even if the contact pressure $p_i(X)$ becomes zero. The modified values are then as follows:

$$T_{mb} = 2B \left\| \int_0^{LL} \tau_m(X)\,dX + \int_{LL}^D m_c[1 - \exp\{-a(j_B + i'_d X)\}\,dX] \right\| \tag{5.118}$$

$$T_{ms} = \int_0^{LL} t_{ms}(X)\,dX + 4H \int_{LL}^D m_c[1 - \exp\{-a(j_B + i'_d X)\}]\,dX \tag{5.119}$$

$$T_{rb} = 2B(R_r + H) \int_0^{\theta r0} m_c[1 - \exp\{-aj_r(\delta)\}]\,d\delta \tag{5.120}$$

$$T_{rs} = 4H(R_r + H) \int_0^{\theta r0} m_c[1 - \exp\{-aj_r(\delta)\}]\cos(\theta_{t0i} - \delta)\,d\delta \tag{5.121}$$

Then, the thrust T_3 can be determined as the sum of the above T_{mb}, T_{ms}, T_{rb}, T_{rs} plus the components T_{fb} given in Eq. (5.84), and T_{fs} given in Eq. (5.86).

By use of the finally values obtained from the above calculations of T_1, T_2, T_3, T_4, P, s_{fi}, s_{ri} and θ_{ti}, each of the energy components given in Eqs. (5.108)–(5.115) and the tractive efficiency E_d given in Eq. (5.116) can be determined. Also, the distributions of the amount of slippage $j(X)$, the contact pressure distribution $p_i(X)$, the shear resistance $\tau_i(X)$ under the track belt and the distribution of the amount of sinkage of the track belt $s_i(X)$ can be calculated. The above calculation can be recursively carried out for various slip ratios i_d,

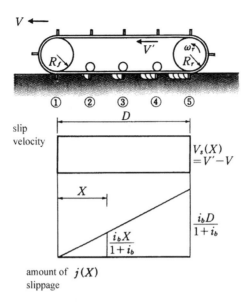

Figure 5.10. Distribution of slip velocity $V_s(X)$ and amount of slippage $j(X)$ under track belt (during braking action).

so that $T_1 \sim T_4 - i_d$ curves, s_{fi}, $s_{ri} - i_d$ curves, e_i, $\theta_{ti} - i_d$ curves, $E_1 \sim E_4 - i_d$ curves and $E_d - i_d$ curve can be drawn graphically. At the same time, the optimum effective driving force T_{4opt}, the optimum slip ratio i_{dopt}, the maximum effective driving force T_m and the corresponding slip ratio i_m can be determined.

5.3   BRAKING STATE ANALYSIS

5.3.1   Amount of vehicle slippage

When a rigid tracked vehicle is running during braking action on a hard terrain at a constant vehicle speed V and a constant circumferential speed of track belt V' ($V > V'$), the slip velocity V_s of a track plate is constant for the whole range of the contact part of the rigid track belt, as shown in Figure 5.10, but the amount of slippage j at each grouser position has a different value.

For a radius of rear sprocket R_r and angular velocity of the braking rear sprocket ω_r, the skid i_b of the vehicle can be defined as:

$$i_b = \frac{V'}{V} - 1 = \frac{R_r\omega_r - V}{V} \tag{5.122}$$

The slip velocity $V_s(X)$ of the track plate at some distance X from the contact point B between the front-idler and the main part of the rigid track belt takes on a negative value. It can be calculated from:

$$V_s(X) = V' - V = R_r\omega_r - V \tag{5.123}$$

Again, the amount of slippage $j(X)$ of a grouser at a distance X can be calculated by integrating the slip velocity V_s from zero to the required time $t = X/R_r\omega_r$ for the movement

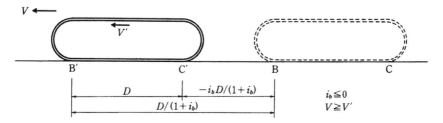

Figure 5.11. Amount of slippage of vehicle during braking action.

of the grouser from the point B to the point X. The calculation is as follows and it has a negative value.

$$j(X) = \int_0^t (R_r \omega_r - V)\, dt = \frac{R_r \omega_r - V}{R_r \omega_R} \cdot X = \frac{i_b}{1 + i_b} \cdot X \qquad (5.124)$$

Yet again, when the track belt moves a distance D around the vehicle itself during braking action, as shown in Figure 5.11, the amount of slippage at the rear end of the track belt $\overline{BC'}$ equals $-i_b D/(1 + i_b)$ while the movement of the vehicle $\overline{BB'}$ becomes $D/(1 + i_b)$, and the required time is given as $D/V' = D/\{V(1 + i_b)\}$.

Thence, the required time t_b when the vehicle has passed totally past the point B can be calculated as:

$$t_b = \frac{D}{V} = (1 + i_b)\frac{D}{V'} \qquad (5.125)$$

The amount of slippage j_s of the soil at the point B in the terrain, associated with the passage of the vehicle, can be calculated as the product of the slip velocity V_s and the transit time of the vehicle t_b as follows:

$$j_s = (V' - V)t_b = i_b V \cdot (1 + i_b)\frac{D}{V'} = i_b D \qquad (5.126)$$

5.3.2   Force balance analysis

Figure 5.12 shows the composite of forces that act on a rigid tracked vehicle when the vehicle is running during braking action at a skid i_b on a horizontal soft terrain. The symbols W, D, B, R_f, R_r, H, G_p, e, h_g, l_d, h_d, V, V', s_{fi}, s_{ri} and p_{fi}, p_{ri} have already been explained in Section 5.2.2.

T_1 is the braking force acting tangentially to the upper part of the rear sprocket, and is equivalent to the braking torque Q divided by the radius of the rear sprocket R_r. T_2 is the compaction resistance that acts horizontally on the front part of the track belt. Its point of application lies at a depth z_i given in Eq. (5.55). T_3 is the drag developed along the rigid track belt through each grouser as the sum of the shear resistance acting on the slip lines that connect each grouser tip. T_4 is the effective braking force that acts horizontally through the point F. P is the ground reaction acting normally on the contact part of the track belt. The eccentricity e_i of P is given in Eqs. (5.57) and (5.58). θ_t is the angle of inclination of the vehicle. It can be calculated from s_{fi} and s_{ri}. The static amount of sinkage $s_{0i}(X)$ of the main part of the track belt can be calculated from Eq. (5.46). The distribution of contact pressure $p_i(X)$ under the track belt can be calculated from Eqs. (5.47) and (5.47).

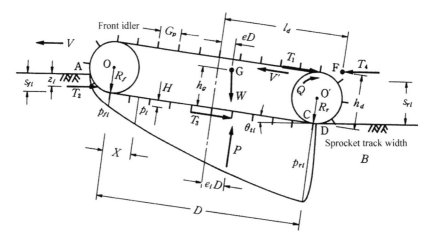

Figure 5.12. Forces acting on a rigid tracked vehicle during braking action.

The equations for force equilibrium in the horizontal and vertical directions and for moment balance around the rear axle have already been established, for driving action, as Eqs. (5.49)∼(5.51) and (5.56). During braking action, all the values of Q, T_1, T_3 and T_4 should be calculated as having negative values.

Thence, the braking force T_1 and the effective braking force T_4 can be expressed as:

$$T_1 = T_3 \tag{5.127}$$

$$T_4 = \frac{T_3}{\cos \theta_{ti}} - W \tan \theta_{ti} - T_2 \tag{5.128}$$

In the above equations, the method of calculation of the drag T_3 and the compaction resistance T_2 will be as outlined in the following two Sections.

5.3.3 Drag

As shown in Figure 5.13 the drag T_3 developed on the contact part of the track belt can be expressed as the sum of a number of components, namely: T_{mb} which is the integral of the shear resistance $\tau_m(X)$; the component T_{ms} which is the integral of the shear resistance $t_{ms}(X)$ which acts on the base and side areas of grousers of the main part of track belt; the component T_{fb} which is the integral of the shear resistance $\tau_f(\theta)$; the component T_{fs} which is the integral of the shear resistance $t_{fs}(\theta)$ which acts on the base and side areas of grousers of the contact part of the front-idler respectively; the component T_{rb} which is the integral of the shear resistance $\tau_r(\delta)$ and the component T_{rs} which is the integral of the shear resistance $t_{rs}(\delta)$ which acts on the base and side areas of grousers of the contact part of the rear sprocket respectively. These shear resistances occur with increases in the amount of sinkage of the track belt. Consequently, the drag T_3 can be expressed as the sum of T_{mb}, T_{ms}, T_{fb}, T_{fs}, T_{rb} and T_{rs} as shown in Eq. (5.59).

(1) Main part of track belt

As a consequence of the main part of the track belt being sloped at an angle equal to that of the angle of inclination of the rigid tracked vehicle θ_{ti} during braking action, the skid i'_b

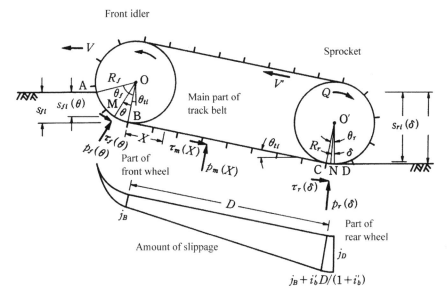

Figure 5.13. Stresses acting on main part of track belt, parts of front and rear wheel and distribution of amount of slippage (during braking action).

needs to be modified – using the value of i_b given in Eq. (5.122) – as follows:

$$i_b' = \frac{V'}{V}\cos\theta_{ti} - 1 = (1 + i_b)\cos\theta_{ti} - 1 \qquad (5.129)$$

Then, the distribution of amount of slippage $j(X)$ on the main part of the rigid track belt can be given for an amount of slippage j_B at point B as:

$$j(X) = j_B + \frac{i_b'}{1 + i_b'}X \qquad (5.130)$$

When j_B equals the maximum amount of slippage j_{fmax} on the contact part of the front-idler, the distribution of shear resistance $\tau_m(X)$ acting on the base area of the grousers of the main part of the track belt can be calculated from the previous Eq. (4.5) as follows:

For the untraction state of $j_q < j(X) < j_p = j_B = j_{fmax}$

$$\tau_m'(X) = \tau_p - k_0\{j_p - j(X)\}^{n_0} \qquad (5.131)$$

For the reciprocal traction state of $j(X) \leq j_q$

$$\tau_m''(X) = -\{m_c + m_f \cdot p_i(X)\}\left[1 - \exp\{-a[j_q - j(X)]\}\right] \qquad (5.132)$$

Here, j_p is the maximum value of the amount of slippage j and τ_p is the corresponding shear resistance to $j = j_p$, and j_q is the value of j when the shear resistance τ becomes zero in the untraction process, i.e.

$$j_q = j_B - \left(\frac{\tau_p}{k_0}\right)^{1/n_0} \qquad (5.133)$$

Again, the distribution of the shear resistance $t_{ms}(X)$ acting on the four sides of the grousers of the main part of track belt can be calculated from the average normal stress F_x/DH acting on the side part of the track belt given in Eq. (5.14) as follows:

For the untraction state of $j_q < j(X) < j_p = j_B = j_{fmax}$

$$t'_{ms}(X) = 4H\left[\tau_p - k_0\{j_p - j(X)\}^{n_0}\right] \tag{5.134}$$

For the reciprocal traction state of $j(X) \leq j_q$

$$t''_{ms}(X) = -4H\left\{m_c + m_f\frac{p_i(X)}{\pi}\cot^{-1}\left(\frac{H}{B}\right)\right\}\left[1 - \exp\{-a[j_q - j(X)]\}\right] \tag{5.135}$$

The shear resistance changes from a positive value to a negative one at $X = DD$ where $j(X)$ becomes j_q in the untraction state. Here, DD can be calculated as follows:

$$DD = -(j_B - j_q)\left(1 + \frac{1}{i'_b}\right) \tag{5.136}$$

In the range $0 \leq X \leq DD$, the amount of slippage $j(X)$ takes on a positive value, while, in the range $DD < X < D$, $j(X)$ takes on a negative value, so that the drag T_{mb} and T_{ms} can be calculated as:

$$T_{mb} = 2B\int_0^{DD}\tau'_m(X)\,dX + 2B\int_{DD}^D\tau''_m(X)\,dX \tag{5.137}$$

$$T_{ms} = \int_0^{DD}t'_{ms}(X)\,dX + \int_{DD}^D t''_{ms}(X)\,dX \tag{5.138}$$

When j_B is less than j_{fmax}, the distribution of shear resistance $\tau(X)$ acting on the main part of the track belt can be calculated for the reciprocal traction state as:

$$\tau''_m(X) = -\{m_c + m_f p_i(X)\}\left\|1 - \exp\left[-a\{j'_q - j(X)\}\right]\right\| \tag{5.139}$$

$$j'_q = j_{f\,max} - (\tau'_p/k_0)^{1/n_0}$$

$$\tau'_p = \{m_c + m_f p_i(X)\}\{1 - \exp(-aj_{f\,max})\}$$

Yet again, the distribution of shear resistance $t_{ms}(X)$ acting on the four sides of the grousers of the main part of track belt can be calculated as:

$$t'''_{ms}(X) = -4H\left\{m_c + m_f\frac{p_i(X)}{\pi}\cot^{-1}\left(\frac{H}{B}\right)\right\} = \left\|1 - \exp\left[1a\{j'_q - j(X)\}\right]\right\| \tag{5.140}$$

In this case, $\tau_m(X)$, $t_{ms}(X)$ take on negative values for the whole range of the main part of track belt. Hence the drag T_{mb} acting on the base area of grousers of the main part of track belt can be calculated as:

$$T_{mb} = 2B\int_0^D\tau'''_m(X)\,dX \tag{5.141}$$

The drag T_{ms} acting on the side part of grousers of the main part of track belt can similarly be calculated as:

$$T_{ms} = 2B\int_0^D t'''_m(X)\,dx \tag{5.142}$$

As the above calculations correspond to $s_{f0} = s_{f0i}$, $s_{r0} = s_{r0i}$ in the cases of Section 5.1.2 (1), (2), (3), $p_i(X)$ should be calculated for each amount of sinkage distribution $s_{0i}(X)$ obtained by substituting $P' = P - P_f$ into $W \cos \theta_{t0}$ as mentioned previously.

Next, let us consider the situation where $s_{f0i} < 0 < H < s_{r0i}$ in Section 5.1.2 (4).

In a similar manner, $s_{0i}(X)$ and $p_i(X)$ can be calculated for $W \cos \theta_{t0} = P$ and the real contact length L of the main part of the track belt can be determined by use of Eq. (5.66). In the range $0 \leq X \leq D - L$, $p_i(X)$ and $\tau_m(X)$ become zero, while, in the range $D - L < X \leq D$, $j(X)$, $\tau'_m(X)$ and $t'_{ms}(X)$ may be given as follows:

$$j(X) = \frac{i'_b}{1 + i''_b}(X - D + L)$$

$$\tau''_m(X) = -\{m_c + m_f p_i(X)\}\,[1 - \exp\{aj(X)\}] \qquad (5.143)$$

$$t''_{ms}(X) = -4H \left\{ m_c + m_f \frac{p_i(X)}{\pi} \cot^{-1}\left(\frac{H}{B}\right) \right\} [1 - \exp\{aj(X)\}] \qquad (5.144)$$

Thence, the drag T_{mb}, T_{ms} can be calculated as follows:

$$T_{mb} = 2B \int_{D-L}^{D} \tau''_m(X)\,dX \qquad (5.145)$$

$$T_{ms} = \int_{D-L}^{D} t''_{ms}(X)\,dX \qquad (5.146)$$

Further, let us consider the case where $s_{f0i} > H > 0 > s_{r0i}$ in Section 5.1.2 (5). Calculating the amount of sinkage $s_{0i}(X)$ and the contact pressure $p_i(X)$ for $W \cos \theta_{t0} = P - P_f$ in the same way as before, the real contact length LL of the main part of the track belt can be calculated from Eq. (5.71).

For the condition $0 \leq X \leq LL$, when j_B equals j_{fmax}, $\tau'_m(X)$ is given by Eq. (5.131) and $t'_{ms}(X)$ is given by Eq. (5.134) when the amount of slippage $j(X)$ given in Eq. (5.130) is in the state of untraction $j_q < j(X) < j_p$, while $\tau''_m(X)$ is given Eq. (5.132) and $t''_{ms}(X)$ is given from Eq. (5.135) when $j(X)$ is less than j_q in the reciprocal traction state. When j_B is less than j_{fmax}, $\tau'''_m(X)$ is given by Eq. (5.139) and $t'''_{ms}(X)$ is given from Eq. (5.140) when $j(X)$ is in the reciprocal traction state.

On the other hand, for $LL < X \leq D$, $\tau''''_m(X)$ for the reciprocal traction of $j(X) < j_q$ can be determined by:

$$\tau''''_m(X) = -m_c \,\|1 - \exp[-a\{j_q - j(X)\}]\| \qquad (5.147)$$

$$t''''_{ms}(X) = -4Hm_c \,\|1 - \exp[-a\{j_q - j(X)\}]\| \qquad (5.148)$$

Thence, when j_B equals j_{fmax}, the value of DD given by Eq. (5.136) is usually in the range of $0 \leq DD \leq LL$, and T_{mb} and T_{ms} can be calculated as follows:

$$T_{mb} = 2B \left\{ \int_0^{DD} \tau'_m(X)\,dX + \int_{DD}^{LL} \tau''_m(X)\,dX + \int_{LL}^{D} \tau''''_m(X)\,dX \right\} \qquad (5.149)$$

$$T_{ms} = \int_0^{DD} t'_{ms}(X)\,dX + \int_{DD}^{LL} t''_{ms}(X)\,dX + \int_{LL}^{D} t''''_{ms}(X)\,dX \qquad (5.150)$$

For $j_B < j_{fmax}$, the shear resistance $\tau(X)$ becomes negative for the whole range of the main part of track belt, so that the drag T_{mb} and T_{ms} can be determined as:

$$T_{mb} = 2B \left\{ \int_0^{LL} \tau_m'''(X)\,dX + \int_{LL}^D \tau_m''''(X)\,dX \right\} \qquad (5.151)$$

$$T_{ms} = \int_0^{LL} t_{ms}'''(X)\,dX + \int_{LL}^D t_{ms}'''(X)\,dX \qquad (5.152)$$

(2) Contact part of the front-idler

The amount of slippage $j_f(\theta)$ at the contact part of the front-idler during braking action can be calculated by use of the entry angle θ_{f0} given in Eq. (5.78) and the angle of inclination of the vehicle θ_{t0i} calculated from Eq. (5.79) by use of the static amount of sinkage of front-idler s_{f0i} and rear sprocket s_{r0i} as follows:

$$j_f(\theta) = (R_f + H) \int_\theta^{\theta_{f0}} \left\{ 1 - \frac{1}{1 + i_b} \cos(\theta + \theta_{t0i}) \right\} d\theta$$

$$= (R_r + H) \left[(\theta_{f0} - \theta) - \frac{1}{1 + i_b} \left\{ \sin(\theta_{f0} + \theta_{t0i}) - \sin(\theta + \theta_{t0i}) \right\} \right] \qquad (5.153)$$

where the amount of slippage at point B, j_B can be given as $j_f(0)$.

The amount of sinkage $s_{f0i}(\theta)$ and the contact pressure $p_{fi}(\theta)$ at the contact part of the front-idler can be calculated from Eq. (5.80) and Eqs. (5.81), (5.82).

Then, the shear resistance of the soil $\tau_f(\theta)$, $t_{fs}(\theta)$ etc. acting on the contact part of the front-idler can be calculated as,

For the traction state of $0 < j_f(\theta) < j_p = j_{fmax}$

$$\tau_f(\theta) = \{m_c + m_f p_{fi}(\theta)\} \left[1 - \exp\{-aj_f(\theta)\} \right] \qquad (5.154)$$

$$t_{fs}(\theta) = 4H \left\{ m_c + m_f \frac{p_{fi}(\theta)}{\pi} \cot^{-1}\left(\frac{H}{B}\right) \right\} \left[1 - \exp\{-aj_f(\theta)\} \right] \qquad (5.155)$$

For the untraction state of $j_q < j_f(\theta) < j_p = j_{fmax}$

$$\tau_f'(\theta) = \tau_p - k_0 \{ j_p - j_f(\theta) \}^{n_0} \qquad (5.156)$$

$$t_{fs}(\theta) = 4H \left[\tau_p - k_0 \{ j_p - j_f(\theta) \}^{n_0} \right] \qquad (5.157)$$

For the reciprocal traction state of $j_f(\theta) < j_q$

$$\tau_f''(\theta) = -\{m_c + m_f \cdot p_{fi}(\theta)\} \left\| 1 - \exp\left[-a\{j_q - j_f(\theta)\}\right] \right\| \qquad (5.158)$$

$$t_{fs}''(\theta) = -4H \left\{ m_c + m_f \frac{p_{fi}(\theta)}{\pi} \cot^{-1}\left(\frac{H}{B}\right) \right\} \left\| 1 - \exp\left[-a\{j_q - j_f(\theta)\}\right] \right\| \qquad (5.159)$$

where j_p is the maximum value of the amount of slippage j and τ_p is the corresponding shear resistance to j_q. The parameter j_q is the value of j when τ becomes zero in the untraction state.

For the case where $\cos\theta_f - 1 < i_b < 0, j_f(\theta)$ equals j_p at $\theta = \theta_{fp}$ and $j_f(\theta)$ equals j_q at $\theta = \theta_{fq}$, then the drag T_{fb} and T_{fs} acting on the base and side areas of the grousers of the contact part of the front idler in the direction of the main straight part of the track belt can be calculated as follows:

$$T_{fb} = 2B(R_r + H) \left[\int_0^{\theta_{fq}} \left\{ \tau_f''(\theta) \cos\theta - p_{fi}(\theta) \sin\theta \right\} d\theta \right.$$

$$\left. + \int_{\theta_{fq}}^{\theta_{fp}} \left\{ \tau_f'(\theta) \cos\theta - p_{fi}(\theta) \sin\theta \right\} d\theta + \int_{\theta_{fp}}^{\theta_{f0}} \left\{ \tau_f(\theta) \cos\theta - p_{fi}(\theta) \sin\theta \right\} d\theta \right]$$

$$(5.160)$$

$$T_{fs} = (R_r + H) \left\{ \int_0^{\theta_{fq}} t_{fs}'' \cos\theta \, d\theta + \int_{\theta_{fq}}^{\theta_{fp}} t_{fs}'(\theta) \cos\theta d\theta + \int_{\theta_{fp}}^{\theta_{f0}} t_{fs}(\theta) \cos\theta \, d\theta \right\} \quad (5.161)$$

Alternately, in the case where $i_b < \cos\theta_f - 1$, the shear resistance $\tau_f(\theta)$ and $t_{fs}(\theta)$ can be given as:

$$\tau_f(\theta) = \left\{ m_c + m_f \cdot p_{fi}(\theta) \right\} \left[1 - \exp\{-aj_f(\theta)\} \right] \qquad (5.162)$$

$$t_{fs}(\theta) = 4H \left\{ m_c + m_f \frac{p_{fi}(\delta)}{\pi} \cot^{-1}\left(\frac{H}{B}\right) \right\} \left[1 - \exp\{-aj_f(\theta)\} \right] \qquad (5.163)$$

and therefore,

$$T_{fb} = 2B(R_f + H) \int_0^{\theta_{f0}} \left\{ \tau_f(\theta) \cos\theta - p_{fi}(\theta) \sin\theta \right\} d\theta \qquad (5.164)$$

$$T_{fs} = (R_f + H) \int_0^{\theta_{f0}} t_{fs}(\theta) \cos\theta \, d\theta \qquad (5.165)$$

The above calculations have been presented for situations corresponding to cases (1), (2), (3) and (5) in Section 5.1.2. For the other case (4), both the values of T_{fb} and T_{fs} become zero.

(3)   *Part of rear sprocket*
The amount of slippage $j_r(\delta)$ of the contact part of the rear sprocket during braking action can be calculated by use of the exit angle $\theta_{r0} = \theta_{r0i}$ and the angle of inclination of the vehicle θ_{t0i}. The inclination angle is given in Eq. (5.79) from the static amount of sinkage of the front-idler s_{f0i} and the rear sprocket s_{r0i}. The overall calculation is then as follows:

$$j_r(\delta) = (R_r + H) \left\{ (\theta_{t0i} - \delta) - \frac{1}{1 + i_b}(\sin\theta_{t0i} - \sin\delta) \right\} + \frac{i_b' D}{1 + i_b'} + j_B \qquad (5.166)$$

The amount of slippage j_D at the point D can be given as $j_r(0)$ from above equation.
    The amount of sinkage $s_{r0i}(\delta)$ and the contact pressure $p_{ri}(\delta)$ of the contact part of the rear sprocket can be calculated through the use of Eqs. (5.88), (5.89) and (5.90).
    Since we know that the amount of slippage $j_r(\delta)$ is usually negative, the shear resistance of the soil $\tau_r(\delta)$ and $t_{rs}(\delta)$ acting on the contact part of the rear sprocket can be calculated as:

$$\tau_r(\delta) = -\left\{ m_c + m_f p_{ri}(\delta) \right\} \left[1 - \exp\{aj_r(\delta)\} \right] \qquad (5.167)$$

$$t_{rs}(\delta) = -4H \left\{ m_c + m_f \frac{p_{fi}(\delta)}{\pi} \cot^{-1}\left(\frac{H}{B}\right) \right\} \left[1 - \exp\{-aj_r(\delta)\} \right] \qquad (5.168)$$

Then, the drag T_{rb} and T_{rs} acting on the base and side areas of the grousers of the contact part of the rear sprocket in the direction of the main part of the track belt $\overline{BC}$ can be calculated as follows:

$$T_{rb} = 2B(R_r + H) \int_0^{\theta_{r0}} \{\tau_r(\delta) \cos(\theta_{t0i} - \delta) + p_{ri}(\delta) \sin(\theta_{t0i} - \delta)\} \, d\delta \tag{5.169}$$

$$T_{rs} = (R_r + H) \int_0^{\theta_{r0}} t_{rs}(\delta) \cos(\theta_{t0i} - \delta) \, d\delta \tag{5.170}$$

The above calculations have been developed for the cases of (1), (2), (3) in Section 5.1.2. For case type (4), T_{rb} and T_{rs} can be calculated by substituting the next amount of slippage $j_r(\delta)$ and the amount of sinkage $s_{r0i}(\delta)$ given in Eq. (5.96) into Eqs. (5.91), (5.92) and Eqs. (5.93), (5.94) respectively.

$$j_r(\delta) = (R_r + H) \left\{ (\theta_{t0i} - \delta) - \frac{1}{1 + i_b}(\sin \theta_{t0i} - \sin \delta) \right\} + \frac{i_b' L}{1 - i_b'} \tag{5.171}$$

On the other hand, for case type (5), T_{rb} and T_{rs} can be calculated by substituting $\tau_r(\delta)$ given Eq. (5.167) and $t_{rs}(\delta)$ given in Eq. (5.103) [which should be modified in consideration of the fact that the term of m_c acts on the contact part of rear sprocket even if the contact pressure $p_{ri}(\delta)$ equals zero] into Eqs. (5.169) and (5.170) respectively.

5.3.4 Compaction resistance

As mentioned in Section 5.2.4, when a rigid tracked vehicle is running on a horizontal soft terrain, the compaction resistance T_2 can be calculated in the manner shown in Eq. (5.103) by use of the amount of sinkage s_{ri} of the rear sprocket. Several methods for the calculation of s_{ri} for various situations will now be presented:

(1) *For the case where* $0 \leq s_{f0i} \leq s_{r0i}$
The total amounts of sinkage s_{fi}, s_{ri} can be calculated by substituting (a) the static amounts of sinkage s_{f0i} and s_{r0i} given in Eq. (5.46) – which are normal to the main part of the rigid track belt, (b) the amounts of slip sinkage s_{fs}, s_{rs} given in the next equations and (c) the angle of inclination of the vehicle θ_{ti} given in Eq. (5.52) into Eqs. (5.97) and (5.98).

The amount of slip sinkage s_{fs} at the point B on the front-idler can be calculated by substituting the contact pressure $p_{fi}(\theta_m)$ and the amount of slippage of the soil j_{fs} into the previous Eq. (4.8) as follows;

$$s_{fs} = c_0 \sum_{m=1}^{M} \{p_{fi}(\theta_m) \cos \theta_m\}^{c_1} \left\{ \left(\frac{m}{M}j_{fs}\right)^{c_2} - \left(\frac{m-1}{M}j_{fs}\right)^{c_2} \right\} \tag{5.172}$$

where

$$\theta_m = \theta_{f0}\left(1 - \frac{m}{M}\right)$$

$$j_{fs} = -(R_f + H)i_b \sin \theta_f \tag{5.173}$$

The amount of slip sinkage s_{rs} at the point C on the rear sprocket can be calculated by substituting the contact pressure $p_i(X)$ and the amount of slippage of the soil j_s into the previous Eq. (4.8) as follows:

$$s_{rs} = s_{fs} + c_0 \sum_{n=1}^{N} \left\{ p_i \left(\frac{nD}{N} \right) \right\}^{c_1} \left[\left(\frac{nj_s}{N} \right)^{c_2} - \left\{ \frac{(n-1)j_s}{N} \right\}^{c_2} \right] \tag{5.174}$$

where

$$J_s = -i_b D$$

(2)   *For the case where $s_{f0i} > s_{r0i} > 0$*
The total amount of sinkage of the rear sprocket s_{ri} following the passage of the front wheel should be larger than the total amount of sinkage s_{fi} of front-idler. Therefore, s_{ri} should be calculated as in the following equation, while s_{fi} is given from Eq. (5.97):

$$s_{ri} = (s_{f0i} + s_{rs}) \cos \theta_{ti} \tag{5.175}$$

(3)   *For the case where $s_{f0i} < 0 < H < s_{r0i}$*
The total amounts of sinkage of front-idler s_{fi} and rear sprocket s_{ri} can be calculated by substituting the amounts of slip sinkage s_{fs} and s_{rs} as shown in the following equations into Eqs. (5.97) and (5.98) respectively:

$$s_{rs} = c_0 \sum_{n=1}^{N} \left\{ p_i \left(\frac{nL}{N} \right) \right\}^{c_1} \left[\left(\frac{nj_s}{N} \right)^{c_2} - \left\{ \frac{(n-1)j_s}{N} \right\}^{c_2} \right] \tag{5.176}$$

where

$$J_s = -i_b L$$

(4)   *For the case where $s_{f0i} > H > 0 > s_{r0i}$*
The total amount of sinkage of the front-idler s_{fi} and the rear sprocket s_{ri} can be calculated by substituting the amount of slip sinkage s_{fs} given in Eq. (5.172) and s_{rs} as shown in the following equation, into Eqs. (5.97) and (5.175) respectively.

$$s_{rs} = s_{fs} + c_0 \sum_{n=1}^{N} \left\{ p_i \left(\frac{nLL}{N} \right) \right\}^{c_1} \left[\left(\frac{nj_s}{N} \right)^{c_2} - \left\{ \frac{(n-1)j_s}{N} \right\}^{c_2} \right] \tag{5.177}$$

where

$$J_s = -i_b LL$$

5.3.5   Energy equilibrium analysis

The effective input energy E_1 supplied by a braking torque to the rear sprocket of a moving tracked-vehicle of necessity equates to the total output energies i.e. to the sum of the component elements namely: the compaction energy E_2 required make a rut under the track belt, the slippage energy E_3 that occurs due to the shear deformation of the soil at the interface between the track belt and the terrain and the effective braking force energy

E_4 which becomes the braking work done on the vehicle. The resultant energy equilibrium equation can then be expressed as:

$$E_1 = E_2 + E_3 + E_4 \tag{5.178}$$

When the track belt moves one contact length of the track belt D relative to the vehicle itself, the vehicle is transferred horizontally to the position of $D \cos \theta_{ti}/(1 + i_b')$. Under these conditions the various energy components can be calculated as in the following equations:

$$E_1 = T_1 D = T_3 D \tag{5.179}$$

$$E_2 = T_2 D \frac{\cos \theta_{ti}}{1 + i_b'} = T_2 D \frac{i}{1 + i_b} \tag{5.180}$$

$$E_3 = T_3 D \frac{i_b'}{1 + i_b'} + WD \frac{1}{1 + i_b'} \sin \theta_{ti} = T_3 D \frac{(1 + i_b) \cos \theta_{ti} - 1}{(1 + i_b) \cos \theta_{ti}} + WD \frac{1}{1 + i_b} \tan \theta_{ti} \tag{5.181}$$

$$E_4 = T_4 D \frac{\cos \theta_{ti}}{1 + i_b'} = T_4 D \frac{1}{1 + i_b} \tag{5.182}$$

Also, each energy component per unit time can be expressed in terms of the vehicle speed V and the circumferential speed of the track belt V' as follows:

$$E_1 = T_1 V' = T_3 V' = T_3 V (1 + i_b) \tag{5.183}$$

$$E_2 = T_2 V' \frac{1}{1 + i_b} = T_2 V \tag{5.184}$$

$$E_3 = T_3 V' \frac{(1 + i_b) \cos \theta_{ti} - 1}{(1 + i_b) \cos \theta_{ti}} + WV' \frac{1}{1 + i_b} \tan \theta_{ti}$$

$$= -T_3 V \left\{ \frac{1}{\cos \theta_{ti}} - (1 + i_b) \right\} + WV \tan \theta_{ti} \tag{5.185}$$

$$E_4 = T_4 V' \frac{1}{1 + i_b} = T_4 V \tag{5.186}$$

5.3.6   Effective braking force

The effective braking force T_4 developed by a rigid tracked vehicle running during braking action can be obtained by substituting the drag T_3, the compaction resistance T_2 and the angle of inclination of the vehicle θ_{ti} into Eq. (5.128). In the $T_4 - i_b$ curve, the 'optimum skid' i_{bopt} is defined as the skid when the effective input energy $|E_1|$ calculated under a constant vehicle speed V takes a maximum value and the 'optimum effective braking force' T_{4opt} is defined as the corresponding effective braking force with i_{bopt}. The 'braking efficiency' E_b is defined as:

$$E_b = \frac{E_4}{E_1} = \frac{VT_4}{V'T_1} = \frac{1}{1 + i_b} \cdot \frac{T_4}{T_1} \tag{5.187}$$

Figure 5.14 shows a flow chart that may be used to calculate the trafficability of a rigid tracked vehicle running on a soft terrain during braking action. At the outset, the vehicle dimensions and the terrain-track system constants are required as input data. After that, the

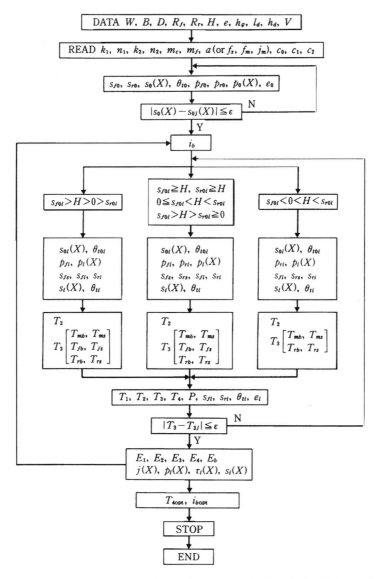

Figure 5.14. Flow chart to calculate braking performance of rigid tracked vehicle running on soft terrain during braking action.

contact pressure distribution p_{f0}, p_{r0} and $p_0(X)$ of the rigid tracked vehicle at rest can be calculated for the static amount of sinkage distribution $s_0(X)$ and the angle of inclination of the vehicle θ_{t0} in each case of $(1) \sim (5)$ as mentioned in Section 5.1.2, and it should be iteratively calculated until the amount of sinkage $s_0(X)$ is strictly determined. Next, the skid i_b during braking action can be computed from $i_b = -1\%$ to -99% successively. The flow system of the calculation can be divided into the following three streams on the basis of the static amounts of sinkage of the front-idler s_{f0i} and the rear sprocket s_{r0i}.

For the cases of $s_{f0i} \geq H$, $s_{r0i} \geq H$; $0 \leq s_{f0i} < H < s_{r0i}$; $s_{f0i} > H > s_{r0i} \geq 0$, the parameters $s_0(X)$ and θ_{t0i} can be calculated from Eqs. (5.46) and (5.79) respectively, and then the factors p_{fi}, p_{ri}, $p_i(X)$ can be calculated from Eqs. (5.47) and (5.48). After that, the amount of slip sinkage s_{fs} and s_{rs} can be calculated from Eqs. (5.172) and (5.174) respectively. Thence, the total amount of sinkage of the front-idler s_{fi} and the rear sprocket s_{ri} can be calculated from Eqs. (5.97) and (5.175) respectively in the case of $0 \leq s_{f0i} \leq s_{r0i}$, and from Eqs. (5.97) and (5.175) respectively in the case of $s_{f0i} > s_{r0i} > 0$. After that, $s_i(X)$ and θ_{ti} can be determined by use of Eqs. (5.102) and (5.52) respectively.

As a next step, the compaction resistance T_2 can be calculated by substituting s_{ri} into Eq. (5.103). The drag T_3 can then be calculated from Eq. (5.59) as the sum of T_{mb} given in Eq. (5.137) or (5.140), T_{ms} given in Eq. (5.138) or (5.141), T_{fb} given in Eq. (5.160), T_{fs} given in Eq. (5.161), T_{rb} given in Eq. (5.169) and T_{rs} given in Eq. (5.170).

Furthermore, the braking force T_1 given in Eq. (5.127), the effective braking force T_4 given in Eq. (5.128) and the ground reaction P given in Eq. (5.54) should be iteratively calculated until the drag T_3 is uniquely determined.

It should be noted, in the process of calculation of the static amount of sinkages of the front-idler and the rear sprocket, that the ground reaction P' acting on the main part of the rigid track belt should be calculated by subtracting the ground reaction P_f acting on the contact part of the front-idler from the above mentioned ground reaction P.

For the case of $s_{f0i} < 0 < H < s_{r0i}$, the factors $s_{0i}(X)$ and θ_{t0i} and the parameters p_{fi} ($=0$), p_{ri}, $p_i(X)$ can be calculated in a similar manner. The amount of slip sinkage of the rear sprocket s_{rs} can be calculated from Eq. (5.176), while that of the front-idler s_{fs} is zero. Thence, the total amount of sinkage of the front-idler s_{fi} and the rear sprocket s_{ri} can be calculated from Eqs. (5.97) and (5.98) respectively and then $s_i(X)$ and θ_{ti} can be determined.

Next, the compaction resistance T_2 can be calculated from the previously obtained amount of sinkage s_{ri}.

The drag T_3 can be calculated as the sum of T_{mb} given in Eq. (5.145), T_{ms} given in Eq. (5.146), T_{rb} given in Eq. (5.92) and T_{rs} given in Eq. (5.94). The braking force T_1, the effective braking force T_4 and the ground reaction P can be iteratively calculated until the drag T_3 is strictly determined.

For the situation where $s_{f0i} > H > 0 > s_{r0i}$, the factors $s_{0i}(X)$ and θ_{t0i} and the parameters p_{fi}, p_{ri} ($=0$), $p_i(X)$ can be calculated in the same way. The amount of slip sinkage of the front-idler s_{fs} can be calculated from Eq. (5.172) and the amount of slip sinkage of the rear sprocket s_{rs} can be calculated from Eq. (5.177). Thence, the total amount of sinkage of the front-idler s_{fi} and the rear sprocket s_{ri} can be calculated from Eqs. (5.97) and (5.175) respectively and then $s_i(X)$ and θ_{ti} can be determined.

Following this, the compaction resistance T_2 can be calculated from the previously obtained amount of sinkage s_{ri}. The drag T_3 can be calculated from Eq. (5.59) as the sum of T_{mb} given in Eq. (5.149) or (5.151), T_{ms} given in Eq. (5.150) or (5.152), T_{fb} given in Eq. (5.160), T_{fs} given in Eq. (5.161), T_{rb} given in Eq. (5.169) and T_{rs} given Eq. (5.170). The braking force T_1 given in Eq. (5.127), the effective braking force T_4 given in Eq. (5.128) and the ground reaction P given in Eq. (5.54) can be iteratively calculated until the drag T_3 is uniquely determined.

For this condition, in the process of calculation of the static amount of sinkage of the front-idler and the rear sprocket, the ground reaction P' acting on the main part of the rigid track belt should be determined by subtracting the ground reaction P_f acting on the contact part of the front-idler from the ground reaction P calculated above.

By use of the output values so obtained i.e. T_1, T_2, T_3, T_4, P, s_{fi}, s_{ri} and θ_{ti}, the various energy components given in Eqs. (5.179) $\sim$ (5.186) as well as the braking efficiency E_b given in Eq. (5.187) can be determined. Also, the distributions of the amount of slippage $j(X)$, contact pressure $p_i(X)$, shear resistance $\tau_i(X)$ under the track belt and the distributions of the amount of sinkage of the track belt $s_i(X)$ can be calculated.

The above calculation can be sequentially carried out for various skids i_b, so that $T_1 \sim T_4 - i_b$ curves, s_{fi}, $s_{ri} - i_b$ curves, e_i, $\theta_{ti} - i_b$ curves, $E_1 \sim E_4 - i_b$ curves and $E_b - i_b$ curve can be plotted. At the same time, the optimum effective braking force T_{4opt}, the optimum skid i_{bopt} can be determined.

5.4   EXPERIMENTAL VALIDATION

The validity of this proposed simulation method for analysis of the trafficability of a rigid tracked vehicle has been confirmed against an experimental test study [9] that used a model rigid tracked vehicle running on a remolded silty loam terrain. The water content of the silty loam was approximately 30% and the cone index of the terrain was approximately 31 kPa. The track model was fitted with standard T shaped steel grousers of track width 20 cm and grouser height $H = 3.2$ cm as shown in Figure 5.15.

The terrain-track system constants obtained from the plate loading test and the plate traction test for the track model plate on the silty loam terrain have previously been given in Table 4.1.

Figure 5.16 shows a plan and side-elevation view of the model rigid-tracked vehicle. The vehicle had a rear wheel drive system comprised of a constant speed motor of 1.5 kW.

The track belt was made of roller chain attached with the same track plate as the model track plate.

As shown in Photo 5.1, this rigid tracked vehicle used a multi-roller system instead of the road roller system which is in general use. The weight of the model rigid tracked vehicle was 3.55 kN. The geometrical details of the vehicle and other specifications are given in Table 5.1.

Figure 5.17 is a schematic of the experimental apparatus. It shows the positions of the test vehicle on the surface of the soil bin during driving and braking action. The size of the steel panel soil-bin was of length 540 cm with a width of 150 cm and a height of 60 cm.

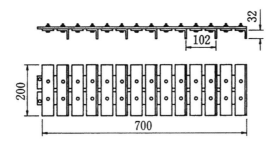

Figure 5.15. Dimensions of track model plate (standard T shaped grouser).

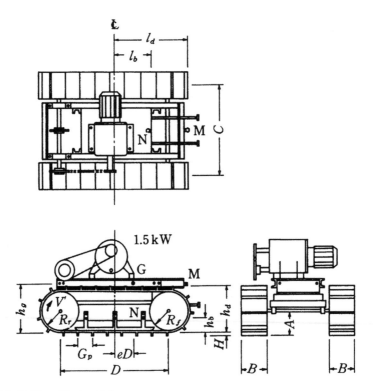

Figure 5.16. Outline of test vehicle.

Photo 5.1. Model rigid tracked vehicle.

Table 5.1. Specification and dimension of rigid tracked vehicle.

Vehicle weight	W	(kN)		3.55
Average contact pressure	p_m	(kPa)		12.5
Height of center of gravity G from bottom of track belt	h_g	(cm)		35.3
Output of motor		(kW)		1.5
Contact length of track belt	D	(cm)		71
Circumferential length of track belt		(cm)		236
Clearance	A	(cm)		13.5
Radius of frontidler	R_f	(cm)		14.8
Radius of rear sprocket	R_r	(cm)		14.8
Track width	B	(cm)		20
Circumferential speed of track belt during driving action	V'	(cm)		9.4
Eccentricity of center of gravity of vehicle	e	(%)		−0.5
Vehicle speed during braking action	V	(cm/s)		9.4
Grouser pitch	G_p	(cm)	5.1 10.2	20.4
Grouser height	H	(cm)		3.2
G_p/H			1.6 3.2	6.4
Distance between application point M of effective tractive effort and central line of vehicle	l_d	(cm)		50.8
Height of application point M of effective tractive effort	h_d	(cm)		32.5
Distance between application point N of effective braking force and central line of vehicle	l_b	(cm)		−17.3
Height of application point N of effective braking force	h_b	(cm)		11.5
Central interval of track belt	C	(cm)		67.2
Thickness of track belt		(cm)		0.3

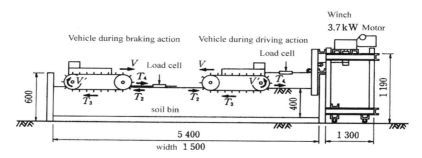

Figure 5.17. Experimental apparatus and vehicles during driving and braking action.

The silty loam test soil was deposited uniformly until a depth of 40 cm was attained. The soil properties were as follows: real specific gravity 2.84, liquid limit 33.2%, plastic limit 21.4%, plastic index of 11.8%, average grain size of 54 μm, coefficient of curvature of 0.31, coefficient of uniformity 6.40, bulk density of 18.6 kN/m³, void ratio of 0.85, cone index of 31 kPa and water content of 29.5 ± 1.0%. The silty loam sample was remolded to have a uniform strength in the vertical direction of the soil bin. After flattening the surface of the test soil with a grader, several trafficability tests were executed.

For this series of tests, the soil sample and the soil bin were the same ones as were used for the plate loading and traction test of the previous track model plate. A tractive apparatus was made to control the vehicle speed and slip ratio or skid during driving or braking action respectively. The power source of the tractive system was a variable speed motor of 3.7 kW. The wire speed out of the winch of the tractive device could be controlled to a range of speeds V from 3.9 cm/s to 23.6 cm/s. The effective tractive effort or the effective braking force was measured using a load cell of maximum capacity 20 kN. The load cell had a sensitivity of 1 N and was connected to the wire and to the vehicle as shown in the diagram.

The experimental method that was employed was as follows. During driving action, the feed speed of the wire, i.e. the vehicle speed V, was controlled to a value somewhat less than the circumferential speed V' of the track belt depending on the various slip ratios i_d employed. For braking action, the vehicle speed V is required be controlled to be larger than the track speed V' depending on the various skids i_b as shown in the diagram. For each test, the driving force or braking force T_1, the effective driving or braking force T_4, the initial amounts of sinkage of the front-idler s_{f0} and the rear sprocket s_{r0} and the steady state amounts of sinkage of the front-idler s_{fi} and the rear sprocket s_{ri} are measured. It is noted that the factor T_1 can be calculated from the measured driving or braking torque divided by the radius of the rear sprocket after subtracting the internal friction between the track belt and the multi-rollers. T_4 can be measured by use of the load cell as mentioned previously. In the test set up this value was recorded automatically.

For a rigid tracked vehicle, Figure 5.18 shows the experimentally derived relationships between T_1, T_4 and slip ratio i for a machine having several track belts of alternate grouser

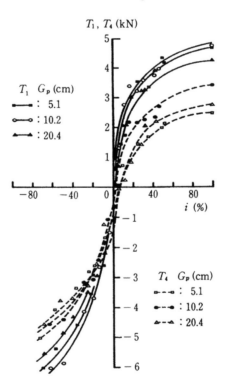

Figure 5.18. Relationship between driving and braking force T_1, effective driving and braking force T_4 and slip ratio or skid i.

pitches $G_p = 5.1$, 10.2 and 20.4 cm respectively. For both the driving and braking state, the larger the slip or skid, the larger are the magnitudes of $|T_1|$, $|T_4|$ that are developed. At the same time, the compaction resistance T_2 increases also with increasing values of $|i|$. It is observed that the factors, $|T_1|$, $|T_4|$ had a maximum value with a grouser pitch $G_p = 10.2$ cm.

The developed driving force T_1 is less than the braking force $|T_1|$ for the same slip ratio and skid $|i|$. This results from the fact that the amount of slippage at the rear end of the track belt during driving action is smaller than that during braking action. As a consequence, the shear resistance is not sufficiently mobilized in the driving state. In quantitative terms this effect can be seen in that the amount of slippage j at the rear end of the track belt of contact length D during braking action is equal to $i_b D/(1 + i_b)$ from Eq. (5.124), while the amount of slippage j during driving action is equal to $i_d D$ from Eq. (5.43).

Figure 5.19 shows the experimentally derived set of relations that prevail between the amount of sinkage of the front-idler s_{fi}, the amount of sinkage of the rear sprocket s_{ri} and the slip ratio i for various values of G_p. During driving action, the amount of sinkage of the rear sprocket s_{ri} increases with the increasing values of slip ratio while the amount of sinkage of the front-idler s_{fi} takes on negative values i.e. the position of the front-idler tends to raise up with slip ratio so that the angle of inclination of the vehicle θ_{ti} increases with i.

During braking action, both the amounts of sinkage s_{fi} and s_{ri} increase with the increasing values of skid $|i_b|$. The amount of sinkage of the rear sprocket s_{ri} is always larger than that of front-idler s_{fi} due to the increasing amount of slippage.

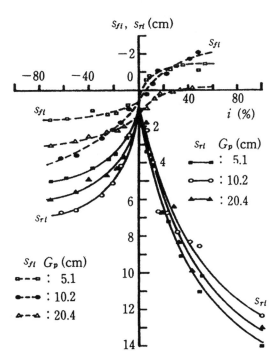

Figure 5.19. Relationship between amounts of sinkage of front idler s_{fi}, rear sprocket s_{ri} and slip ratio or skid i.

Photo 5.2. Rut of track belt during driving action ($i_d = 49\%$, $G_p = 10.2$ cm).

Photo 5.3. Rut of track belt during braking action ($i_b = -50\%$, $G_p = 10.2$ cm).

The amount of sinkage of the rear sprocket i.e. the rut depth of the vehicle s_{ri} during driving action takes on larger values than those during braking action for the same slip ratio and skid $|i|$. This is because the amount of slippage j_s of soil in the passage of the vehicle during driving action equates to $i_d D/(1 - i_d)$ as shown in Eq. (5.45) rather than that during braking action of $i_b D$ as shown in Eq. (5.126).

These phenomena can be verified from a detailed study of the shape of the rut depth as shown in Photo 5.2.

This photo was taken during driving action at a slip ratio of $i_d = 49\%$. Photo 5.3 was taken during braking action at a skid of $i_b = -50\%$.

Moreover, the rut depth s_{ri} during driving action took its maximum value at $G_p = 5.1$ cm. The nearest follower was $G_p = 20.4$ with 10.2 cm next. The rut depth s_{ri} during braking action reached a maximum value when $G_p = 10.2$ cm followed by $G_p = 20.4$ and 5.1 cm respectively.

Figure 5.20 shows the experimental relations that developed between the angle of inclination of the vehicle θ_{ti} and the slip ratio during driving action or with skid during braking action i. The tests were for a machine operating in a running steady-state. The values for

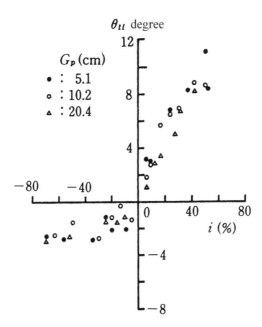

Figure 5.20. Relationship between angle of inclination of vehicle θ_{ti} and slip ratio or skid i.

Photo 5.4. Model rigid tracked vehicle running during driving action ($i_d = 49\%$, $G_p = 5.1$ cm).

the angle θ_{ti} are larger during driving action than during braking action, so that the position of the front-idler rises up beyond the surface of terrain and consequently the eccentricity of ground reaction exceeds 1/6.

This process is confirmed in Photo 5.4. The photo was taken during driving action. As an example of a simulation analysis of the trafficability of a rigid tracked vehicle [10],

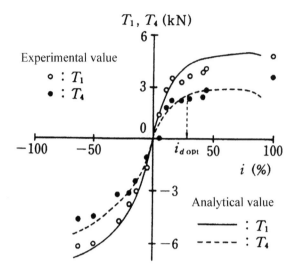

Figure 5.21. Relationship between driving and braking force T_1, effective driving and braking force T_4 and slip ratio or skid.

the tractive and the braking performance of a model rigid tracked vehicle having the spec-ification as shown in Table 5.1 on the terrain having the terrain-track system constants as shown in the previous Table 4.1 was simulated for both driving and braking action. The vehi-cle dimensions and the terrain-track system constants for the grouser pitch of $G_p = 10.2$ cm were used as input data in the flow chart as shown in Figures 5.9 and 5.14. Here, there was no need to consider the size effect of the track belt on the terrain-track system constants because the contact area of the model track belt as shown in Figure 5.15 was almost the same value as that of the model rigid tracked vehicle.

Figure 5.21 shows the comparison between the measured and theoretical value in the relationship between the driving and braking force T_1, the effective driving and braking force T_4 and the slip ratio or skid i. Both the measured and theoretical results agree very well with each other. Also, the analytical results are well verified by the experimental data. For all values of slip ratio or skid i, the difference between the driving force T_1 and the effective driving force T_4 became larger than the difference between the braking force $|T_1|$ and the effective braking force $|T_4|$. This is because the compaction resistance T_2 calculated from the amount of slip sinkage due to the amount of slippage during driving action becomes larger than that during braking action, and the component of the ground reaction P in the direction of T_4, e.g. $P \sin \theta_{ti}$ during driving action increases with the corresponding increment of the angle of inclination of vehicle as shown in Figure 5.20 rather than that during braking action. That is, the sum of T_2 and $P \sin \theta_{ti}$ in Eq. (5.49) during driving action becomes larger than that during braking action.

Figure 5.22 shows the comparison between the measured and theoretical value in the relationship between the amount of sinkage of front-idler s_{fi}, rear sprocket s_{ri} and slip ratio during driving action or skid during braking action i. In these cases, the analytical results agree well with the experimental results. The amount of sinkage of rear sprocket s_{ri} during driving action became almost twice in comparison with that during braking action. This is because the amount of slippage j_s of soil in the passage of the vehicle during driving action

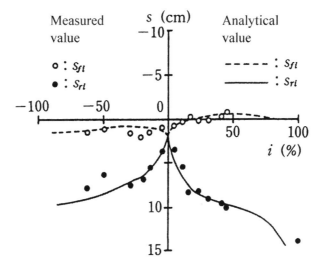

Figure 5.22. Relationship between amounts of sinkage of front idler s_{fi}, rear sprocket s_{ri} and slip ratio or skid i.

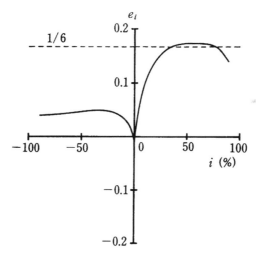

Figure 5.23. Relationship between eccentricity of ground reaction e_i and slip ratio or skid i.

becomes larger than that during braking action, as mentioned previously. Moreover, it can be explained theoretically that the amount of slip sinkage increases remarkably when the slip ratio during driving action approaches 100%.

Figure 5.23 shows the analytical relationship between the eccentricity e_i of the ground reaction and slip ratio during driving action or skid during braking action i respectively. The eccentricity during driving action increases with the increment of slip ratio and there is some ranges showing $e_i > 1/6$, while the eccentricity during braking actions becomes less than 0.05. These phenomena correspond well with the tendencies of the amounts of sinkage of the front-idler and rear sprocket.

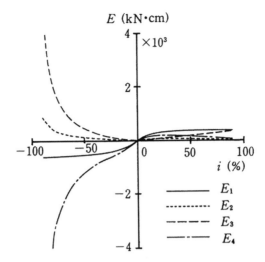

Figure 5.24. Relations between energy components E and slip ratio or skid i during passage of track belt D.

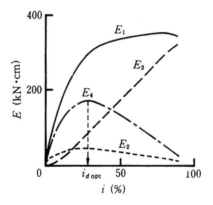

Figure 5.25. Relationship between energy components E and slip ratio i during driving action (extended diagram).

Figure 5.24 shows the analytical relationships pertaining to each of the energy components E_1, E_2, E_3 and E_4 during the passage of the track belt of the main contact length D around the vehicle and slip ratio i. Comparing each energy component during driving action, E_2, E_3 and E_4 during braking action increases remarkably with increasing values of skid $|i|$. This is because E_2, E_3 and E_4 take on an infinite value in Eqs. (5.180) and (5.182) when the skid i_b approaches -100%, while E_1 approaches a constant value in Eq. (5.179). In the calculation of E_2, E_3 and E_4 during braking action, the moving distance of the vehicle becomes $D/(1+i_b)$ which is also very large comparing with the moving distance of vehicle $D(1-i_d)$ during driving action. Figure 5.25 is an enlarged diagram of $E_1 \sim E_4 - i$ relations during driving action of Figure 5.24. E_3 increases parabolically with the increases in the slip ratio i_d, while E_1 approaches a constant value. On the other hand, E_2 and E_4 takes a maximum value at some slip ratio respectively. A tracked vehicle can develop a maximum

Table 5.2. Dimensions of rigid tracked vehicle.

Vehicle weight	W	(cm)	40.0
Average contact pressure	p_m	(kPa)	23.0
Height of center of gravity G from bottom of track belt	h_g	(cm)	50
Contact length of track belt	D	(cm)	170
Width of track belt	B	(cm)	50
Radius of frontidler	R_f	(cm)	25
Radius of rear sprocket	R_r	(cm)	25
Vehicle speed during braking action	V	(cm/s)	100
Grouser height	H	(cm)	6.5
Grouser pitch	G_p	(cm)	14.6
Eccentricity of center of gravity	e		0.00
Distance between application point of effective tractive effort and braking force and central line of vehicle	l	(cm)	120
Height of application point of effective tractive effort and braking force	h	(cm)	30
Circumferential speed of track belt during driving action	V'	(cm/s)	100

tractive work at an optimum slip ratio i_{dopt} when E_4 takes a maximum value. In this case, the effective driving force T_{4opt} is 2.44 kN at the optimum slip ratio of $i_{opt} = 28\%$, and the maximum effective driving energy E_4 is 166.6 kNcm.

The usefulness of the application of the simulation analytical method as presented in the previous Section can be verified from the experimental results of the model rigid tracked vehicle. Using this method, it is possible to understand properly the tractive and braking performance of a given bulldozer running on any kind of soft terrain. In future, it will be shown that this simulation analytical method is valid and useful for design and production of new bulldozers for running on soft terrain.

5.5 ANALYTICAL EXAMPLE

5.5.1 Pavement road

In the following section, we present some simulation analysis results for a rigid tracked vehicle of weight 40 kN running during driving and braking action on a flat concrete pavement. The simulation has been carried out by use of the flow chart given in Figures 5.9 and 5.14. The general properties of the vehicle are as shown in Table 5.2. The terrain-track system constants for the equilateral trapezoidal rubber grousers of base length $L = 3$ cm and for the concrete pavement system are as given in Figure 4.6 and the previous Table 4.3.

For the simulation, it is assumed that the pavement road behaves as an elastic material, that the terrain-track system constant k_1 equals 9.8×10^4 N/cm$^{n_1+1}$, and that $n_1 = 1$, $k_2 = n_2 = 0$ and $c_0 = c_1 = c_2 = 0$. The size effect of the constants f_s, f_m, j_m is not considered.

In terms of the outcomes of running the simulation are concerned, let us now examine the results for the driving state. We will look at the braking state results a little later.

Figure 5.26 is a plot of the relations between the driving force T_1, the effective driving force T_4 and the slip ratio i_d. For this case, T_4 works out to be equal to T_1 since the compaction resistance T_2 becomes zero for the negligibly small amount of sinkage of the track belt. T_4 shows an almost constant value, after it takes an early maximum value of 41.25 kN at an optimum slip ratio of $i_{opt} = 1\%$.

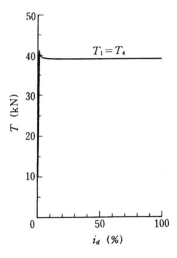

Figure 5.26. Relationship between driving force T_1, effective driving force T_4 and slip ratio i_d during driving action (concrete pavement road).

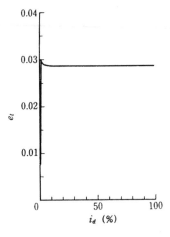

Figure 5.27. Relationship between eccentricity of ground reaction and slip ratio during driving action (concrete pavement road).

Figure 5.27 shows the relationship between the eccentricity e_i of the ground reaction and the slip ratio i_d. The eccentricity e_i takes up an almost constant value, after it passes a peak value of 0.0303 at $i_{dopt} = 1\%$. Figure 5.28 shows the relationships between the individual energy components E_1, E_2, E_3, E_4 and slip ratio i_d. E_1 stays at an essentially constant value due to the constant circumferential speed of the track belt V'. The component E_2 does not occur so no land locomotion resistance develops. The component E_3 increases almost linearly with increasing values of slip ratio i_d. E_4 decreases almost linearly with i_d to a value of zero at $i_d = 100\%$ after it takes a peak value of 4083 kNcm/s at $i_{dopt} = 1\%$.

Additionally, as shown in Figure 5.29, the tractive efficiency E_d becomes very high and peaks at a maximum value of 98.9% at $i_{dopt} = 1\%$. After this maximum it decreases gradually and almost linearly with i_d to a value of zero at $i_d = 100\%$.

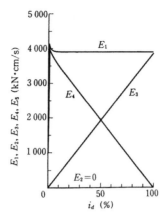

Figure 5.28. Relationship between energy components E_1 to E_4 and slip ratio during driving action (concrete pavement road).

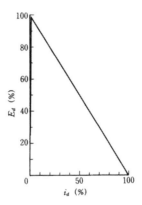

Figure 5.29. Relationship between tractive efficiency E_4 and slip ratio i_4 during driving action (concrete pavement road).

Figure 5.30 shows the longitudinal distribution of normal pressure for $i_d = 10$, 20 and 30% and the longitudinal distributions of shear resistance for $i_d = 1$, 5, 10, 20 and 30%. The normal pressure distribution increases monotonically toward the rear end of the track belt i.e. the rear sprocket and it increases irrespective of the slip ratio. On the other hand, the shear resistance distribution varies quite dramatically with the amount of slippage for small magnitudes of slip ratio i_d. It tends, though, to show a constant Hump type curve having a peak value just under the front-idler for large magnitudes of slip ratio i_d.

Following these considerations of the driving state, the simulation results for the braking state are now presented.

Figure 5.31 shows the relations between the braking force T_1, the effective braking force T_4 and the skid i_b. In this case, T_4 also computes to be equal to T_1 since the compaction resistance T_2 becomes zero for the negligibly small amount of sinkage of the track belt that occur in this case. The braking force T_4 takes on an almost constant value, after it reaches an early minimum value of -41.78 kN at an optimum skid $i_{bopt} = -1\%$.

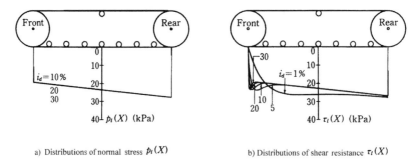

a) Distributions of normal stress $p_t(X)$ b) Distributions of shear resistance $\tau_t(X)$

Figure 5.30. Distribution of contact pressure during driving state (concrete pavement road).

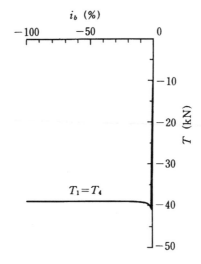

Figure 5.31. Relationship between braking force T_1 and effective braking force T_4 and skid i_b during braking action (concrete pavement road).

Figure 5.32 shows the relationship between the eccentricity e_i and the skid i_b. The eccentricity e_i is essentially constant value, after it takes a peak value of -0.0307 at $i_{bopt} = -1\%$.

Figure 5.33 shows the relationship between the energy components E_1, E_2, E_3, E_4 and skid i_b. $|E_1|$ decreases almost linearly with increasing values of $|i_b|$ to a value of zero at $i_b = -100\%$. Prior to this however it has peaked at a maximum value of $4139\,\text{kNcm/s}$ at $i_{bopt} = -1\%$. E_2 does not occur in this case and as a consequence no land locomotion resistance occurs. The component E_3 increases almost linearly with the increasing values of skid $|i_b|$. E_4 shows a constant value due to the constant vehicle speed V. As a consequence of these conditions, the braking efficiency E_b becomes very high as shown in Eq. (5.187) and increases hyperbolically with increasing values of $|i_b|$.

Figure 5.34 shows the longitudinal distributions of normal stress at $i_b = -10, -20\%$ and the longitudinal of shear resistance at $i_b = -1, -5, -10$ and -20%. The normal pressure distribution decreases monotonically toward the rear end of the track belt e.g. the rear sprocket and it decreases irrespective of the value of the skid.

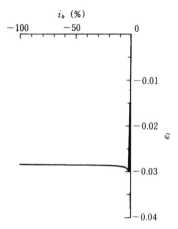

Figure 5.32. Relationship between eccentricity e_i of ground reaction and skid i_b during braking action (concrete pavement road).

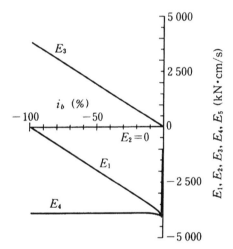

Figure 5.33. Relationship between energy components $E_1 \sim E_4$ and skid i_b during braking action (concrete pavement road).

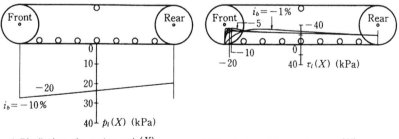

a) Distributions of normal stress $p_i(X)$ b) Distributions of shear resistance $\tau_i(X)$

Figure 5.34. Distribution of contact pressure during braking state (concrete pavement road).

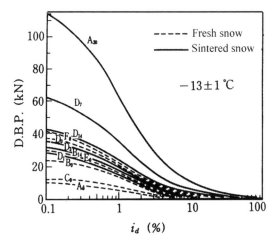

Figure 5.35. Relationship between effective tractive effort (D.B.P.) of over-snow vehicle and slip ratio i_d running on various snow covered terrains [14].

On the other hand, the shear resistance distribution varies quite dramatically with amount of slippage for small magnitudes of skid $|i_b|$. However, it tends to show a constant Hump type curve having a peak value just under the front-idler for the large magnitudes of skid $|i_b|$.

5.5.2   Snow covered terrain

In this section, the results of an energy analysis [11] for a tracked over-snow vehicle running on a snow covered terrain composed of a newly fallen dry snow and sintered snow under low temperature of -13 °C will be presented. The findings are based on a study [12] of rectangular plate loading tests and a series [13] of vane cone tests carried out on a snow covered terrain.

When the engine power of a tracked over-snow vehicle is large enough to develop a driving torque at the rear sprocket greater than the maximum thrust occurring at the interface between the track belt and the snow covered ground, the effective driving force of the tracked over-snow vehicle takes a maximum value at a slip ratio of zero and after that it decreases rapidly with increasing slip ratio, as shown in Figure 5.35 [14]. Here, the compression and shear deformation characteristics of each snow sample of $A_0 \sim D_{14}$ exhibit a rigid plastic behaviour under low temperature, as presented previously in Sections 1.3.2 and 1.3.3. In this case, the maximum contact pressure of the rigid track belt of contact length 285 cm and track width of 74 cm is calculated to be 33.9 kPa.

Additionally, the effective driving force of a tracked over-snow vehicle running on various kinds of snow covered terrain composed of different snow materials decreases linearly with increases in the compressive deformation energy of snow for any depths of deposited snow. That is, as shown in Figure 5.36, the effective driving force D.B.P. (kN) can be expressed [15] as a function of the maximum driving force T (kN) transmitted from the engine, the contact length of the track belt D (m) and the compressive deformation energy E_D (kNm) of the snow. That is:

$$\text{D.B.P.} \approx T - E_D/D \tag{5.188}$$

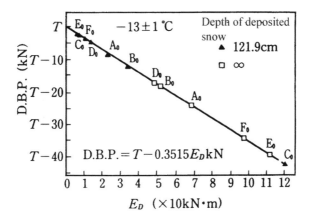

Figure 5.36. Relationship between effective tractive effort D.B.P. of over-snow vehicle and compressive deformation energy E_D for various snow covered terrains (T: Maximum driving force).

where E_D is the penetration work of the rigid track belt into the snow until the contact pressure reaches a maximum for the track belt.

In general, E_D increases with increasing depth of snow and the effective tractive effort i.e. the effective driving force of the tracked over-snow vehicle decreases with increases in snow-thickness.

Next, let us present some results for an energy analysis [16] of a tracked over-snow vehicle running on a snow covered terrain of a depth of 10 cm which is composed of wet snow at the temperature of 0 °C. The analysis is based on the results of circular plate loading tests and ring shear test results that have been carried out on the snowy terrain. In this case, the trafficability of a tracked over-snow vehicle of the contact length 350 cm, the width 50 cm, with an eccentricity in center of gravity of the vehicle of zero and an average contact pressure of 8 kPa has been calculated for machine operations under driving action. The results, as given in Figure 5.37 show the relationships that exist between the various energy components E_1, E_2, E_3, E_4 and the slip ratio i_d. The components E_1 and E_4 increase parabolically with increasing values of i_d. E_2 shows an almost constant value. E_3 increases slightly with i_d.

Figure 5.38 displays the calculated relationship between the effective tractive effort D.B.P. and the slip ratio i_d for several average contact pressure p_{av}. This data is for a tracked over-snow vehicle having the same vehicle dimensions as before. As the average contact pressure increases, the deposited snow is compressed so that the strength of the snow becomes high. Thence, the maximum effective tractive effort increases and the corresponding slip ratio decreases at the same time. From this diagram, it can be suggested that there is an optimum average contact pressure that maximizes the maximum effective tractive effort for each average contact pressure.

Figure 5.39 indicates that there is an optimum average contact pressure that maximizes the maximum effective tractive effort for each value of eccentricity in the center of gravity of a tracked over-snow vehicle. This phenomenon has also been observed in connection with studies of super weak clayey soil [17]. That is, the maximum draw-bar pull Max. D.B.P. increases with increase in the average contact pressure, but decreases gradually with the increase in the land locomotion resistance that accompanies the increasing compressive deformation energy of the deposited snow that develops after the average contact pressure

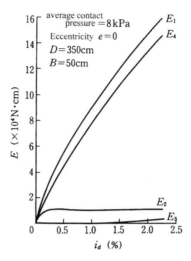

Figure 5.37. Relationship between energy components $E_1 \sim E_4$ and slip ratio i_d (wet snow, depth of deposited snow $= 10$ cm).

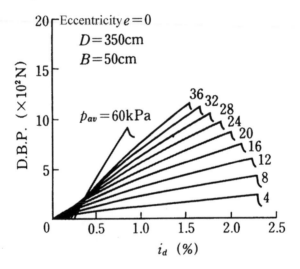

Figure 5.38. Relationship between effective tractive effort D.B.P. of over-snow vehicle and slip ratio i_d for various average contact pressures p_{av} (wet snow, depth of deposited snow 10 cm).

exceeds the optimum average contact pressure. In this case, the optimum average contact pressure occurs somewhere between 27.4 and 42.1 kPa and the maximum value of the maximum effective tractive effort i.e. the maximum draw-bar pull Max. D.B.P. decreases with increase in the eccentricity e of the center of gravity of the vehicle.

5.6 SUMMARY

In this chapter, we have set out to analyse the behaviour of the simplest of the tracked vehicle systems – namely that of the rigid-track types – using the model-track method.

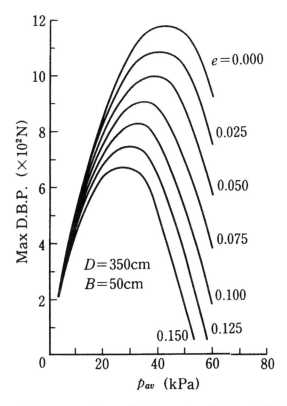

Figure 5.39. Relationship between maximum effective tractive effort Max. D.B.P. and average contact pressure for various eccentricities of center of gravity of vehicle (wet snow, depth of deposited snow 10 cm).

By use of computer simulation and through the use of shaped-plate traction test data (as discussed in Chapter 4) the behaviour of tracked vehicles operating under a number of different physical circumstances can be predicted and thence analysed.

In this chapter a mix of analytical and empirical methods and parameters have been used to predict:

– The ground pressures under an immobile tracked vehicle as a function of different inclination angles and extents of track sinkage.
– The amount of sinkage, rut depth and drawbar pull that a specific design of tracked vehicle will produce while it is driving forward or under braking.
– The over-ground speed of the vehicle.

By systematically varying a machine's design parameters and/or the ground conditions, trafficability and mobility studies of specific mechanical engineering configurations can be explored.

REFERENCES

1. The Japanese Geotechnical Society (1982). *Handbook of Soil Engineering*. pp. 303–343. (In Japanese).

2. Terzaghi, K. (1943). *Theoretical Soil Mechanics.* pp. 118–143. John Wiley & Sons.
3. Akai, K. (1975). *Soil Mechanics.* pp. 204–209. Asakura Press. (In Japanese).
4. Muro, T. (1991). Optimum Track Belt Tension and Height of Application Forces of a Bulldozer Running on Weak Terrain. *J. of Terramechanics, Vol. 28, No. 2/3*, pp. 243–268.
5. Hata, S. (1987). *Construction Machinery.* pp. 76–91. Kajima Press.
6. Muro, T. (1989). Stress and Slippage Distributions under Track Belt Running on a Weak Terrain. *Soils and Foundations, Vol. 29, No. 3*, pp. 115–126.
7. Muro, T. (1989). Tractive Performance of a Bulldozer Running on Weak Ground. *J. of Terramechanics, Vol. 26, No. 3/4*, pp. 249–273.
8. Muro, T. (1990). Control System of the Optimum Height of Application Forces for a Bulldozer Running on Weak Terrain. *Proc. of the 1st Symposium on Construction Robotics in Japan*, pp.197–206. JSCE et al. (In Japanese).
9. Muro, T., Omoto, K. & Futamura, M. (1988). Traffic Performance of a Bulldozer Running on a Weak Terrain-Vehicle Model Test. *J. of JSCE No. 397/VI-9*, pp. 151–157. (In Japanese).
10. Muro, T., Omoto, K. & Nagira, A. (1989). Traffic Performance of a Bulldozer Running on a Weak Terrain – Energy Analysis. *J. of JSCE, No. 403/VI-10*, pp. 1103–110. (In Japanese).
11. Muro, T. & Yong, R.N. (1980). On Trafficability of Tracked Oversnow Vehicle – Energy Analysis for Track Motion on Snow Covered Terrain. *Journal of Japanese Society of Snow and Ice, Vol. 42, No. 2*, pp. 93–100. (In Japanese).
12. Muro, T. & Yong, R.N. (1980). Rectangular Plate Loading Test on Snow Mobility of Tracked Oversnow – *Journal of Japanese Society of Snow and Ice, Vol. 42, No. 1*, pp. 25–32. (In Japanese).
13. Muro, T. & Yong, R.N. (1980). On Drawbar Pull of Tracked Oversnow Vehicle. *Journal of Japanese Society of Snow and Ice, Vol. 42, No. 2*, pp. 101–109. (In Japanese).
14. Yong, R.N. & Muro, T. (1981) Prediction of Drawbar-Pull of Tracked Over-Snow Vehicle. *Proc. of 7th Int. Conf. of ISTVS, Calgary, Canada. Vol. 3*, pp. 1119–1149.
15. Muro, T. (1981). Shallow Snow Performance of Tracked Vehicle. *Soils and Foundations, Vol. 24, No. 1*, pp. 63–76.
16. Muro, T. & Enoki, M. (1983). Trafficability of Tracked Vehicle on Super Weak Ground. *Memoirs of the Faculty of Engineering, Ehime University, Vol. X, No. 2*, pp. 329–338. (In Japanese).

EXERCISES

(1) Suppose that a bulldozer is running during driving action on a sandy terrain at a slip ratio of $i_d = 30\%$. The contact length of the bulldozer D is 3.5 m. The radius of the rear sprocket R_r is 50 cm and the angular velocity of rotation ω_r is π rad/s. Calculate the slip velocity V_s between the track and terrain, and the amount of slippage $j(D)$ at the rear end of the bulldozer.

(2) The bulldozer given in previous problem (1) has passed an arbitrary point X in the running lane. Calculate the transit time t_d and the amount of slippage j_s at the point X.

(3) Imagine a bulldozer of weight $W = 100$ (kN) running during driving action on a soft terrain at a slip ratio of $i_d = 20\%$. The contact length D of the bulldozer is 4 (m) and the width of the track B is 45 (cm). It is observed that the amount of sinkage at the front-idler s_f was 3 (cm) and that at the rear sprocket s_r was 5 (cm). Calculate the thrust T_{mb} acting on the base area of the grousers of the main part of track belt, assuming that the distribution of the ground reaction is trapezoidal and the amount of eccentricity of the ground reaction e_i is 0.05. Assume that the terrain-track system constant $m_c = 50$ (kPa), $m_f = 0.458$ and $a = 0.244$ (1/cm) and that the amount of slippage j_B at the position of the front-idler can be neglected.

(4) Calculate the compaction resistance T_2 and the effective tractive effort T_4 of the bulldozer given in previous problem (3). Assume the height of grouser $H = 2.0$ cm, the

coefficient of sinkage $k_1 = 0.763$ N/cm$^{n_1+1}$, $k_2 = 1.491$ N/cm$^{n_2+1}$, and the indices of sinkage $n_1 = 0.866$, $n_2 = 0.842$.

(5) Suppose that the bulldozer given in previous problem (3) is running at a speed $V = 100$ cm/s. Confirm the fact that at a slip ratio $i_d = 20\%$, the effective input energy E_1 equals the sum of the compaction energy E_2, the slippage energy E_3 and the effective tractive effort energy E_4.

(6) A tractor of contact length $D = 2.5$ m is running during braking action on a sandy terrain at a skid value $i_b = -20\%$. The circumferential speed of the rear sprocket $R_r \omega_r$ is 50π cm/s. Calculate the slip velocity V_s between the tractor and terrain, and the amount of slippage $j(D)$ at the rear end of the tractor.

(7) The tractor given in previous problem (6) has passed an arbitrary point X in the running lane. Calculate the transit time t_b and the amount of slippage j_s at the point X.

(8) Imagine that a bulldozer of weight $W = 100$ (kN) is running during braking action on a soft terrain at a skid of $i_b = -20\%$. The contact length D of the bulldozer is 400 (cm) and the width of track B is 45 (cm). It is observed that the amount of sinkage of the front-idler s_f is 3.5 (cm) and that of rear sprocket s_r is 4.5 (cm). Calculate the drag T_{mb} acting on the base area of the grousers of the main part of the track belt, assuming that the distribution of the ground reaction is trapezoidal and the amount of eccentricity e_i of the ground reaction is 0.02. Assume that the terrain-track system constants are $m_c = 50$ kPa, $m_f = 0.655$ and $a = 0.258$ (1/cm) and that the amount of slippage j_B at the position of front-idler can be neglected.

(9) Calculate the effective braking force T_4 and the compaction resistance T_2 of the bulldozer given in problem (8). Assume the height of grouser $H = 2.0$ cm, the coefficient of sinkage $k_1 = 0.763$ N/cm$^{n_1+1}$, $k_2 = 1.491$ N/cm$^{n_1+2}$ and the indices of sinkage $n_1 = 0.866$, $n_2 = 0.842$.

(10) The bulldozer given in previous problem (8) is running at the speed of $V = 100$ cm/s. Confirm the fact that, at the skid $i_b = -20\%$, the effective input energy E_1 is equal to the sum of the compaction energy E_2, the slippage energy E_3 and the effective braking force energy E_4.

Chapter 6

Land Locomotion Mechanics of Flexible-Track Vehicles

The common 'flexible track belt' system that is mounted on many bulldozers or tractors comes in two general mechanical engineering styles. In the first style, only up-down movements of a track plate that spans between mutually connected road rollers can occur. This arrangement is as shown in Figure 6.1. In the second structural configuration or style, up-down and side-side movements of road rollers that are mutually connected by hinges, springs or torsion bars can occur. This second style is shown in Figure 6.2. In general, the flexible track system is appropriate for construction work that must take place on or over terrains that are geometrically rough or which are full of ups and downs.

In the sections that follow, the land locomotion mechanics and some methods of analysis of a flexible tracked vehicle in which the axles of the road rollers are mutually connected, will be presented. Consideration will also be given to the behaviour of the flexible track belt under conditions of decreasing initial track belt tension. The analytical methods so developed may also be extendable to the case where vertical and lateral movement of the road rollers can occur – such as happens in systems which have various styles of suspension system.

6.1   FORCE SYSTEM AND ENERGY EQUILIBRIUM ANALYSIS

Figure 6.3 shows the composite of forces that act on a flexibly tracked vehicle when the vehicle is climbing a slope under driving action or is descending a slope under braking

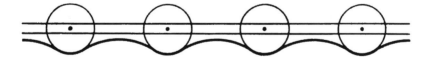

Figure 6.1. Flexible track belt under mutually-connected road rollers.

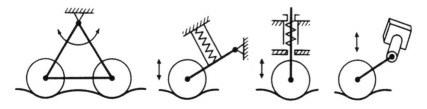

Figure 6.2. Flexible track belt under road rollers connected by hinge, spring or torsion bar.

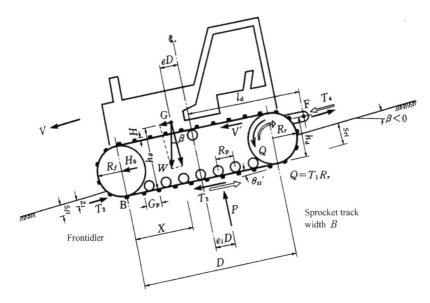

Figure 6.3. Vehicle dimension and several forces acting on several forces during driving (←) or braking (⇐) action on soft terrain of slope ange β.

action. The slope in this case is defined to have a slope angle β. The symbols and nomenclature in the diagram are as have been previously introduced in Sections 5.2.2 and 5.3.2. A positive driving torque or a negative braking torque Q acts on the rear sprocket and a corresponding positive driving force or negative braking force T_1 acts on the lower contact part of the track belt or on the upper suspended part of the track belt. The compaction resistance T_2 is assumed to act on the front contact part of the track belt at a depth z_i and to operate in a direction parallel to the sloped terrain surface during both driving and braking actions. The depth z_i can be calculated by carrying out a moment balance around the axle of the rear sprocket when an effective braking force i.e. a pure rolling resistance $T_4 = -T_2$ is applied to the point F under conditions where the angle of inclination of the vehicle is θ'_{ti}, the amount of sinkage of the rear sprocket is s'_{ri} and where $T_1 = T_3 = W \sin(\theta'_{ti} + \beta)$ and $P = W \cos(\theta'_{ti} + \beta)$ for the pure rolling state. The depth can then be calculated as:

$$z_i = s'_{ri} - \left[T_2 \left\{ R_r + (h_d - R_r) \cos \theta'_{ti} - \left(l_d - \frac{D}{2} \right) \sin \theta'_{ti} \right\} \right.$$

$$\left. - W \left\{ h_g \sin (\theta'_{ti} + \beta) - D \left(\frac{1}{2} - e \right) \cos(\theta'_{ti} + \beta) \right\} \right] /$$

$$\left\{ T_2 + W \cos (\theta'_{ti} + \beta)/ \tan \theta'_{ti} \right\} \qquad (6.1)$$

In this equation θ'_{ti} can be calculated from the amounts of sinkage of the front-idler s'_{fi} and the rear sprocket s'_{ri} via the following equation.

$$\theta'_{ti} = \sin^{-1} \left(\frac{s'_{ri} - s'_{fi}}{D} \right) \qquad (6.2)$$

The positive thrust or negative drag T_3, which can be calculated as the sum of the shear resistances that develop at the interface between the soil and track belt, is assumed to act at the tips of grousers and operate in a direction parallel to the main contact part of the flexible track belt. The positive effective driving force or negative effective braking force T_4 is assumed to act at a point F in a direction parallel to the sloped terrain surface. Generally, the eccentricity e_i of the ground reaction P can be calculated from a moment balance around the axle of the rear sprocket during driving or braking action as follows:

$$e_i = \frac{1}{2} + \frac{1}{PD}\left[-T_2(R_r - s'_{ri} + z_i) + T_4\left\{ (h_d - R_r)\cos\theta'_{ti} - \left(l_d - \frac{D}{2} \right)\sin\theta'_{ti} \right\} \right.$$

$$\left. + W\left\{ (h_g - R_r)\sin(\theta'_{ti} + \beta) - D\left(\frac{1}{2} - e \right)\cos(\theta'_{ti} + \beta) \right\} \right] \tag{6.3}$$

When a flexibly tracked vehicle is at rest $i_d = i_b = 0$, $T_1 = T_3 = W\sin(\theta'_{ti} + \beta)$, $T_2 = T_4$, $Q = WR\sin(\theta'_{ti} + \beta)$ and the eccentricity e_0 of the ground reaction P can be given as follows:

$$e_0 = e + \frac{h_g - R_r}{D}\tan(\theta'_{ti} + \beta) \tag{6.4}$$

Considering force equilibrium conditions between T_1, T_2, T_3, T_4, P and W in the parallel and normal directions to the surface of the sloping terrain and from a moment balance around the axle of the rear sprocket, the following relations are obtained.

$$T_2 + T_4 = T_3\cos\theta'_{ti} - P\sin\theta'_{ti} - W\sin\beta \tag{6.5}$$

$$W\cos\beta = T_3\sin\theta'_{ti} + P\cos\theta'_{ti} \tag{6.6}$$

$$Q = T_1 R_r = T_3 R_r \tag{6.7}$$

Thence, the driving or braking force T_1, the effective driving or braking force T_4 and the ground reaction P can be calculated as:

$$T_1 = T_3 \tag{6.8}$$

$$T_4 = \frac{T_3}{\cos\theta'_{ti}} - \frac{W\sin(\theta'_{ti} + \beta)}{\cos\theta'_{ti}} - T_2 \tag{6.9}$$

$$P = \frac{W\cos\beta}{\cos\theta'_{ti}} - T_3\tan\theta'_{ti} \tag{6.10}$$

Here, under the conditions of $i_d = i_b = 0$, the above equations are satisfied for $T_2 = T_4 = 0$, $T_3 = W\sin(\theta'_{t0} + \beta)$, $\theta'_{ti} = \theta'_{t0}$ and $P = W\cos(\theta'_{t0} + \beta)$.

Next, let us consider equilibrium in the energy domain. The effective input energy E_1 supplied by the driving or braking torque acting on the rear sprocket must equal the total output energies i.e. the sum of the energy components comprise of the compaction energy E_2 that is required to make a rut under the flexible track belt, the slippage energy E_3 that develops in the shear deformation of the soil beneath the flexible track belt, the effective driving or braking force energy E_4 and the potential energy E_5. This yields the following balance equation.

$$E_1 = E_2 + E_3 + E_4 + E_5 \tag{6.11}$$

In addition, the energy expenditure or consumption per unit of time can be worked out from the vehicle speed V and the circumferential speed V' of the flexible track belt as follows:

(1) *During driving action*

$$E_1 = T_1 V' = T_3 V' = T_3 \frac{V}{1 - i_d} \tag{6.12}$$

$$E_2 = T_2 V'(1 - i_d) = T_2 V \tag{6.13}$$

$$E_3 = T_3 \left(1 - \frac{1 - i_d}{\cos \theta'_{ti}} \right) V' + WV'(1 - i_d) \tan \theta'_{ti} \cos \beta$$

$$= T_3 \left(\frac{1}{1 - i_d} - \frac{1}{\cos \theta'_{ti}} \right) V + WV \tan \theta'_{ti} \cos \beta \tag{6.14}$$

$$E_4 = T_4 V'(1 - i_d) = T_4 V \tag{6.15}$$

$$E_5 = WV'(1 - i_d) \sin \beta = WV \sin \beta \tag{6.16}$$

(2) *During braking action*

$$E_1 = T_1 V' = T_3 V' = T_3 V(1 + i_b) \tag{6.17}$$

$$E_2 = T_2 \frac{V'}{1 + i_b} = T_2 V \tag{6.18}$$

$$E_3 = T_3 \frac{(1 + i_b) \cos \theta'_{ti} - 1}{(1 + i_b) \cos \theta'_{ti}} V' + WV' \frac{1}{1 + i_b} \tan \theta'_{ti} \cos \beta$$

$$= T_3 \left\{ (1 + i_b) - \frac{1}{\cos \theta'_{ti}} \right\} V + WV \tan \theta'_{ti} \cos \beta \tag{6.19}$$

$$E_4 = T_4 \frac{V'}{1 + i_b} = T_4 V \tag{6.20}$$

$$E_5 = W \frac{V'}{1 + i_b} \sin \beta = WV \sin \beta \tag{6.21}$$

6.2   FLEXIBLE DEFORMATION OF A TRACK BELT

As shown in Figure 6.4, the flexible deformation characteristics of the track belt of a flexibly tracked vehicle depend mainly on the track belt tension T_0, the ground reaction F_p, and the shear resistance F_s acting on the interface between the soil and the track belt. For a section $\overline{CC_m}$ of the flexible track belt as shown in the diagram, one can set-up some force balance equations between the forces T_0 and F_s acting in the longitudinal direction to the row of road rollers and between F_p acting in the normal direction to the row of road rollers. This process yields the following relationships:

$$T_0 + F_s = T_0 + \Delta T_0 \tag{6.22}$$

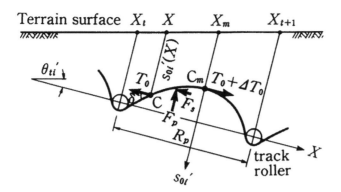

Figure 6.4. Forces acting on a flexible track belt.

$$F_p = T_0 \tan \delta = T_0 \frac{ds'_{oi}(X)}{dX} \tag{6.23}$$

In these equations F_s is the integral of the shear resistance of the soil acting along the X axis i.e. along the line of the row of road rollers from the coordinate X to X_m corresponding to the points C to C_m, and it equals the increase of track belt tension. F_p is computable as the integral of the normal stresses that act on the track belt as one passes from X to X_m. The angle of deflection of the track belt at an arbitrary point C is designated as δ. The point C_m is taken to be situated at the position where the amount of deflection of the track belt takes on a maximum value and where the angle δ takes a value of zero. The parameter $s'_{0i}(X)$ is the static amount of sinkage of the flexible track belt at a distance X from the bottom-dead-center B of the front-idler where it contacts with the main straight part of the track belt.

The force $F_p = F(X)$ can be approximated as the integral of the normal contact pressure $p_i(X)$ from X to X_m assuming a rigid track belt as follows:

$$F(X) = B \int_X^{X_m} p_i(X) \, dX \tag{6.24}$$

where B is the width of the track belt.

The force $T_0 = T_0(X)$ can be calculated as the sum of the initial track belt tension H_0 and the thrust $H_m(X)$ i.e. the integral of the shear resistance $\tau_i(X)$ of soil acting on the interface between the track belt and the terrain over the range $X = 0$ to $X = X$, as follows:

$$T_0(X) = H_0 + H_m(X) = H_0 + B \int_0^X \tau_i(X) \, dX \tag{6.25}$$

Then, the static amount of sinkage $s'_{0i}(X)$ can be calculated by integrating Eq. (6.23). The result is:

$$s'_{0i}(X) = \int_X^{X_m} \frac{F(X)}{T_0(X)} dX + s'_{0i}(X_m) - \{s'_{0i}(D) - s'_{0i}(0)\} \frac{X_m - X}{D} \tag{6.26}$$

Further, the distributions of the normal stress $p_i(X)$ that acts under a flexible track belt can be sub-analysed as follows:

For the situation where $0 \leq s'_{0i}(X) \leq H$

$$p'_i(X) = k_1 \{s'_{0i}(X)\}^{n_1} \tag{6.27}$$

For the situation where $s'_{0i}(X) > H$

$$p'_i(X) = k_1 H^{n_1} + k_2 \{s'_{0i}(X) - H\}^{n_2} \tag{6.28}$$

In the unloading phase that occurs after the passage of a road roller, the distribution of the normal contact pressure $p_i(X)$ at a distance X where $s'_{0i}(X)$ becomes less than the peak value of static amount of sinkage $s'_{0i}(X_t)$ at the coordinate X_t of each road roller [1] can be sub-analysed as follows:

For the situation where $0 \leq s'_{0i}(X_t) \leq H$

$$p'_i(X) = k_1 \{s'_{0i}(X_t)\}^{n_1} - k_3 \{s'_{0i}(X_t) - s'_{0i}(X)\}^{n_3} \tag{6.29}$$

For the situation where $s'_{0i}(X_t) > H$

$$p'_i(X) = k_1 H^{n_1} + k_2 \{s'_{0i}(X_t) - H\}^{n_2} - k_4 \{s'_{0i}(X_t) - s'_{0i}(X)\}^{n_4} \tag{6.30}$$

where the normal stress at the front-idler p'_{fi} and at the rear sprocket p'_{ri} are given as $p'_i(0)$ and $p'_i(D)$ respectively.

Additionally, as mentioned in the previous Sections 5.2.6 and 5.3.6, the static amount of sinkage $s'_{0i}(X)$ should be calculated by taking into account the fact that the integrated value of $p'_i(X)$ acting on the main straight part of the track belt must always be equal to the value of the real ground reaction $P' = P - P_f$. Thence, the total amount of sinkage of the front-idler s'_{fi} and the rear sprocket s'_{ri} in the case of $0 \leq s'_{f0i} \leq s'_{r0i}$ can be expressed as:

$$s'_{fi} = (s'_{f0i} + s'_{fs}) \cos \theta'_{ti} \tag{6.31}$$

$$s'_{ri} = (s'_{r0i} + s'_{rs}) \cos \theta'_{ti} \tag{6.32}$$

$$\theta'_{ti} = \sin^{-1} \left(\frac{s'_{ri} - s'_{fi}}{D} \right) \tag{6.33}$$

Under the conditions where $s'_{f0i} > s'_{r0i} > 0$, the total amount of sinkage of the rear sprocket s'_{ri} following after the passage of the front-idler should be greater than the total amount of sinkage of the front-idler s'_{fi}. Thus, s'_{ri} given in the above Eq. (6.32) needs to be modified as follows:

$$s'_{ri} = (s'_{f0i} + s'_{rs}) \cos \theta'_{ti} \tag{6.34}$$

Here, $s'_{f0i} = s'_{0i}(0)$ and $s'_{r0i} = s'_{0i}(D)$, and s'_{fs} and s'_{rs} are the amounts of slip sinkage of the front-idler and the rear sprocket respectively.

The amount of sinkage $s_i'(X_t)$ at each contact point X_t of the front-idler, the road rollers and the rear sprocket to the main straight part of the track belt are required to lie on the same straight line, so that the following identity can be established.

$$s_i'(X_t) = s_{fi}' + (s_{ri}' - s_{fi}')\frac{X_t}{D} \tag{6.35}$$

Further, the total amount of sinkage $s_i'(X)$ of the flexible track belt at a distance X can be calculated as follows:

$$s_i'(X) = s_{0i}'(X) + s_i'(X_t) - s_{0i}'(X_t)$$
$$+ \left[\{s_i'(X_{t+1}) - s_{0i}'(X_{t+1})\} - \{s_i'(X_t) - s_{0i}'(X_t)\}\right]\frac{X - X_t}{R_p} \tag{6.36}$$

where R_p is the spacing interval of each road roller.

The thrust or drag T_3 can be calculated, as will be discussed later, as the sum of the components T_{mb}, T_{mb}', T_{mb}'', T_{ms}, T_{ms}', T_{ms}'' that act on the main contact part of the track belt, the components T_{fb}, T_{fs} that act on the front-idler, and the elements T_{rb}, T_{rb}', T_{rs} and T_{rs}' that act on the rear sprocket. Also, the compaction resistance T_2 can be calculated for a rut depth $s_{ri}' > H$ in the manner already mentioned in conjunction with the previous Eq. (5.103). Thus,

$$T_2 = 2B\left[\int_0^H k_1 z^{n_1} dz + \int_H^{s_{ri}'} \{k_1 H^{n_1} + k_2(z - H)^{n_2}\} dz\right]$$
$$= \frac{2k_1 B}{n_1 + 1}H^{n_1+1} + 2k_1 BH^{n_1}(s_{ri}' - H) + \frac{2k_2 B}{n_2 + 1}(s_{ri}' - H)^{n_2+1} \tag{6.37}$$

6.3   SIMULATION ANALYSIS

Figure 6.5 shows a flow chart that can be used to calculate the traffic performance of a flexible tracked vehicle ascending a soft sloping terrain of slope angle β during driving action and also for descending down a terrain during braking action.

As input data to the flow chart the complete dimensions of the vehicle, its speed and the shape of the terrain are required i.e. the vehicle weight W, the track width B, the track contact length D, the radius of the front-idler R_f, the radius of the rear sprocket R_r, the radius of the road roller R_m, the grouser height H, the eccentricity of the center of gravity G of the vehicle e, the height of the center of gravity G from the bottom surface of the track belt h_g, the distance of the point of application of tractive effort F from the center line of the bulldozer l_d, the height of the point F from the bottom surface of the track belt h_d, the initial track belt tension H_0, the slope angle of the terrain β, as well as the vehicle speed V and the circumferential speed of the track belt V'. Then as a second form of input data, the terrain-track system constants k_1, n_1, k_2, n_2, k_3, n_3, k_4, n_4 obtained from track-model-plate loading and unloading test results, and the other terrain-track system constants m_c, m_f, a, k_0, n_0 are required. The constants f_m, f_s, j_m and K_1, K_2, j_m as well as c_0, c_1, c_2 that can be obtained from the plate traction, untraction and reciprocal traction test results as mentioned previously in Chapter 4 are also needed. After these factors are available, the contact pressure distributions p_{f0}, p_{r0} and $p_0(X)$ for a tracked vehicle at rest can be

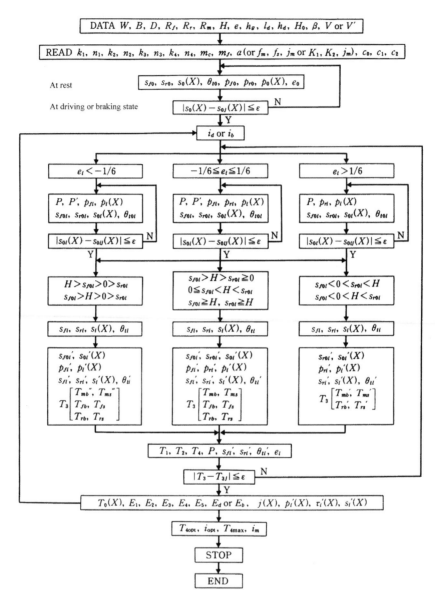

Figure 6.5. Flow chart to analyse the performance of flexible tracked vehicles during driving and braking action.

calculated (assuming a rigid track belt) for the static amounts of sinkage of front-idler s_{f0} and rear sprocket s_{r0}, and the linear distribution of static amount of sinkage $s_0(X)$ under the rigid track belt, and the angle of inclination of vehicle θ_{t0} in each case of (1) ~ (5) as mentioned in the previous Section 5.1.2, and it should be iteratively calculated until the amount of sinkage $s_0(X)$ is uniquely determined.

Next, the slip ratio i_d during driving action or the skid i_b during braking action can be calculated by recursion. The system of calculation may be divided into three different calculation streams on the basis of the value of the eccentricity of ground reaction e_i (just calculated) for ranges of $e_i < -1/6$, $-1/6 \leq e_i \leq 1/6$ and $e_i > 1/6$.

The real ground reaction acting on the main part of the track belt P' can be calculated from the apparent ground reaction P given in the previous Eq. (5.54) by subtracting the ground reaction acting on the contact part of the front-idler P_f given in the previous Eq. (5.117). The linear distribution of the static amount of sinkage s_{f0i}, s_{r0i}, $s_{0i}(X)$ given in the previous Eq. (5.46) and the angle of inclination of the vehicle θ_{t0i} given in the previous Eq. (5.79) can be calculated from the real ground reaction P', and these should be calculated until the contact pressure distribution $p_i(X)$ given in the previous Eqs. (5.47) and (5.48) is uniquely determined. The amounts of slip sinkage of the front-idler s_{fs} and the rear sprocket s_{rs} can be calculated from the previous Eqs. (5.99) and (5.101) respectively. Then, the total amounts of sinkage of the front-idler s_{fi} and the rear sprocket s_{ri} can be calculated for the situation where $0 \leq s_{f0i} \leq s_{r0i}$ from the previous Eqs. (5.97) and (5.98) respectively, and for the situation where $s_{f0i} > s_{r0i} > 0$ from the previous Eqs. (5.97) and (5.104). Thence, the distribution of the amount of sinkage $s_i(X)$ and the angle of inclination of the vehicle θ_{ti} can be determined from the previous Eqs. (5.102) and (5.52) respectively.

As a next step in the calculation process, the static amounts of sinkage of the front-idler s'_{f0i} and the rear sprocket s'_{r0i} of the flexible track belt and the distribution of the static amount of sinkage $s'_{0i}(X)$ can be worked out from Eq. (6.26).

Then, the contact pressure p'_{fi} at the front-idler and p'_{ri} at the rear sprocket, and the contact pressure distribution $p'_i(X)$ acting on the flexible track belt can be calculated from Eqs. (6.27) ~ (6.30).

The total amounts of sinkage s'_{fi} at the front-idler and s'_{ri} at the rear sprocket can be calculated in Eqs. (6.31), (6.32) or (6.34) from the amounts of slip sinkage s'_{fs} at the front-idler and s'_{rs} at the rear sprocket as mentioned later. Also, the angle of inclination of the vehicle θ'_{ti} can be calculated from Eq. (6.33). Finally, the distribution of the total amount of sinkage $s'_i(X)$ can be calculated from Eq. (6.36).

The thrust or drag T_3 can be calculated as the sum of T_{mb} and T_{ms} i.e. the integral of the shear resistance given as a function of the distribution of normal contact pressure $p'_i(X)$ which acts on the interface between the soil and the base area and the side parts of the grousers in the main part of the flexible track belt respectively, T_{fb} and T_{fs} i.e. the integral of the shear resistance of the soil acting on the base area and the side parts of grousers in the contact parts of the front-idler respectively, and T_{rb}, T_{rs} i.e. the integral of the shear resistance of the soil acting on the base area and the side parts of the grousers in the contact parts of the rear sprocket, respectively.

The driving or braking force T_1, the compaction resistance T_2 given in Eq. (6.37), the effective driving or braking force T_4 given in Eq. (6.9), the ground reaction P given in Eq. (6.10), the total amount of sinkage s'_{fi}, s'_{ri} and the angle of inclination of the vehicle θ'_{ti} should be iteratively calculated until the thrust or drag T_3 is uniquely determined. After that, the relations between the distribution of track tension $T_0(X)$, each energy component E_1, E_2, E_3, E_4, and E_5, the distribution of amount of slippage $j(X)$ under track belt, the distribution of normal stress $p'_i(X)$ and shear resistance $\tau'_i(X)$, the distribution of amount of deflection of the track belt and the slip ratio i can be calculated and these relationships can be drawn graphically. Further, the various functions T_1, $T_4 - i$, s'_{fi}, $s'_{ri} - i$, $\theta'_{ti} - i$,

$e_i - i$ and the tractive or braking power efficiency E_d, $E_b - i$ can be plotted graphically by use of a micro-computer. Finally, the optimum effective driving or braking force T_{4opt} at the optimum slip ratio or skid i_{opt} and the maximum effective driving or braking force at the slip ratio or skid i_m can be simultaneously determined.

6.3.1   At driving state

In this section, several methods for the calculation of a number of factors i.e. the amount of slip sinkage s'_{fs}, s'_{rs} during driving action, the total amounts of sinkage of the front-idler s'_{fi} and of the rear sprocket s'_{ri} and the thrust T_3 developed under the flexible track belt, will be presented in detail. These factors relate to the flow chart of Figure 6.5.

The amount of slip sinkage s'_{fs} at the front-idler as shown in Eq. (6.31) can be calculated by substituting the contact pressure distribution $p_f(\theta_m)$ acting on the contact part of the front-idler and the amount of slippage of soil j_{fs} during driving action into the previous Eq. (4.8) as follows:

$$s'_{fs} = c_0 \sum_{m=1}^{M} \{p_f(\theta_m) \cos \theta_m\}^{c_1} \left\{ \left(\frac{m}{M} j_{fs}\right)^{c_2} - \left(\frac{m-1}{M} j_{fs}\right)^{c_2} \right\}$$ (6.38)

where

$$\theta_m = \theta_f \left(1 - \frac{m}{M}\right)$$

$$\theta_f = \cos^{-1}\left(\cos \theta'_{ti} - \frac{s'_{fi}}{R_f + H}\right) - \theta'_{ti}$$

$$p_f(\theta_m) = k_1 \{s(\theta_m)\}^{n_1} \quad (0 \le s(\theta_m) \le H)$$

$$= k_1 H^{n_1} + k_2 \{s(\theta_m) - H\}^{n_2} \quad (s(\theta_m) > H)$$

$$s(\theta_m) = \frac{(R_f + H)\{\cos(\theta_m + \theta'_{ti}) - \cos(\theta_f + \theta'_{ti})\}}{\cos(\theta_m + \theta'_{ti})}$$

$$j_{fs} = (R_f + H) \sin \theta_f \frac{i_d}{1 - i_d}$$ (6.39)

The amount of slip sinkage s'_{rs} at the rear sprocket as expressed in Eq. (6.32) can be calculated by substituting the contact pressure distribution $p'_i(X)$ acting on the main part of the flexible track belt and the amount of slippage of soil j_s during driving action into the previous Eq. (4.8) as follows:

$$s'_{rs} = s'_{fs} + c_0 \sum_{n=1}^{N} \left\{p'_i\left(\frac{n}{N}D\right)\right\}^{c_1} \left\{ \left(\frac{n}{N} j_s\right)^{c_2} - \left(\frac{n-1}{N} j_s\right)^{c_2} \right\}$$ (6.40)

where

$$j_s = \frac{i'_d D}{1 - i'_d}$$

$$i'_d = 1 - \frac{1 - i_d}{\cos \theta'_{ti}}$$

As a next step in the process, a force balance equation involving the thrust T_3 during driving action can be set up as follows:

$$T_3 = T_{mb} + T_{ms} + T_{fb} + T_{fs} + T_{rb} + T_{rs} \qquad (6.41)$$

In this equation, T_{mb} and T_{ms} are components of thrust that act along the base area and the side parts of the grousers of the main part of the flexible track belt, respectively. These thrust components can be calculated as follows:

$$T_{mb} = 2B \int_0^D \{m_c + m_f p'_i(X)\} \left[1 - \exp\{-a(j_w + j_f + i'_d X)\} \right] dX \qquad (6.42)$$

$$T_{ms} = 4H \int_0^D \left\{ m_c + m_f \frac{p'_i(X)}{\pi} \cot^{-1}\left(\frac{H}{B}\right) \right\}$$
$$\times \left[1 - \exp\{-a(j_w + j_f + i'_d X)\} \right] dX \qquad (6.43)$$

$$j_f = (R_f + H) \left[\theta_f - (1 - i_d) \{ \sin(\theta_f + \theta'_{ti}) - \sin \theta'_{ti} \} \right] \qquad (6.44)$$

The elements T_{fb} and T_{fs} are those components of thrust that act on the base area and the side parts of the grousers of the contact part of the front-idler respectively. The magnitude of these elements can be calculated as follows:

$$T_{fb} = 2B(R_f + H) \int_0^{\theta_f} \| \{ m_c + m_f p_f(\theta) \}$$
$$\times \left[1 - \exp\{-a j_f(\theta)\} \right] \cos \theta - p_f(\theta) \sin \theta \| \, d\theta \qquad (6.45)$$

$$T_{fs} = 4H(R_f + H) \int_0^{\theta_f} \left\{ m_c + m_f \frac{p_f(\theta)}{\pi} \cot^{-1}\left(\frac{H}{B}\right) \right\}$$
$$\times \left[1 - \exp\{-a j_f(\theta)\} \right] \cos \theta \, d\theta \qquad (6.46)$$

$$j_f(\theta) = (R_f + H) \left[(\theta_f - \theta) - (1 - i_d) \{ \sin(\theta_f + \theta'_{ti}) - \sin(\theta + \theta'_{ti}) \} \right] + j_w \qquad (6.47)$$

The further factors T_{rb} and T_{rs} i.e. the components of thrust acting on the base area and the side parts of the grousers of the contact part of the rear sprocket, respectively can be calculated as follows:

$$T_{rb} = 2B(R_r + H) \int_0^{\theta_r} \| \{ m_c + m_f \cdot p_r(\delta) \} \left[1 - \exp\{-a j_r(\delta)\} \right]$$
$$\times \cos(\theta'_{ti} - \delta) + p_r(\delta) \sin(\theta'_{ti} - \delta) \| \, d\delta \qquad (6.48)$$

$$T_{rs} = 4H(R_r + H) \int_0^{\theta_r} \left\{ m_c + m_f \frac{p_r(\delta)}{\pi} \cot^{-1}\left(\frac{H}{B}\right) \right\}$$
$$\times \left[1 - \exp\{-a j_r(\delta)\} \right] \cos(\theta'_{ti} - \delta) \, d\delta \qquad (6.49)$$

$$j_r(\delta) = (R_r + H) \{ (\theta'_{ti} - \delta) - (1 - i_d)(\sin \theta'_{ti} - \sin \delta) \} + i'_d D + j_f + j_w \qquad (6.50)$$

In the above equations, j_w is the amount of slippage that occurs along the sloping terrain due to the component of the vehicle weight $W \sin(\theta'_{ti} + \beta)$ that acts on the main contact part of the flexible track belt. On a sloping terrain, the amount of slippage j_w is distributed uniformly along the terrain-track interface at a slip value $i_d = 0$ or at a skid value $i_b = 0$. The value of j_w can be calculated from the following force balance equation which can be set-up between the vehicle weight component and the integral of the shear resistance i.e.

$$W \sin(\theta'_{ti} + \beta) = \pm 2B \int_0^D \{m_c + m_f p'_i(X)\} \{1 - \exp(\pm a j_w)\} \, dX \tag{6.51}$$

and therefore the slippage

$$j_w = \pm \frac{1}{a} \log \left[1 \mp \frac{W \sin(\theta'_{ti} + \beta)}{2B \int_0^D \{m_c + m_f p'_i(X)\} \, dX} \right] \tag{6.52}$$

is obtained. Here, j_w takes on a positive value for values $\theta'_{ti} + \beta > 0$ in relation to the upper sign and takes on a negative value for $\theta'_{ti} + \beta < 0$ in relation to the lower sign.

Further, when the eccentricity e_i of the ground reaction lies outside the middle-third of the main contact part of the flexible track belt during driving action, the thrusts T'_{mb} and T'_{ms} acting on the base area and side parts of the grousers of the main contact part of the track belt and the thrusts T'_{rb} and T'_{rs} acting on the base area and side parts of the grousers of the contact part of the rear sprocket force for $e_i > 1/6$, and the thrusts T''_{mb} and T''_{ms} acting on the base area and side parts of the grousers of the main contact part of the track belt for $e_i < -1/6$ can be calculated in the same way as has already been mentioned in Chapter 5. These procedures are shown in the following equations:

For $s'_{f0i} < 0 < H < s'_{r0i}$

$$T'_{mb} = 2B \int_{D-L}^D \{m_c + m_f p'_i(X)\} \times \|1 - \exp[-a\{j_w + i'_d(X - D + L)\}]\| \, dX$$

$$L = \frac{s'_{r0i}}{s'_{r0i} - s'_{f0i}} D \tag{6.53}$$

$$T'_{ms} = 4H \int_{D-L}^D \left\{ m_c + m_f \frac{p'_i(X)}{\pi} \cot^{-1} \left(\frac{H}{B} \right) \right\}$$
$$\times \|1 - \exp[-a\{j_w + i'_d(X - D + L)\}]\| \, dX \tag{6.54}$$

The values of T'_{rb} and T'_{rs} can be calculated by substituting the amount of slippage $j_r(\delta)$ given in the next equation into Eqs. (6.48) and (6.49).

$$j_r(\delta) = (R_r + H)\{(\theta'_{ti} - \delta) - (1 - i_d)(\sin\theta'_{ti} - \sin\delta)\} + i'_d L + j_w \tag{6.55}$$

For $s'_{f0i} > H > 0 > s'_{r0i}$

$$T'_{mb} = 2B \int_0^{LL} \{m_c + m_f p'_i(X)\} \left[1 - \exp\{-a(j_w + j_f + i'_d X)\}\right] dX$$

$$LL = \frac{s'_{f0i}}{s'_{f0i} + s'_{r0i}} D \tag{6.56}$$

$$T''_{ms} = 4H \int_0^{LL} \left\{m_c + m_f \frac{p'_i(X)}{\pi} \cot^{-1}\left(\frac{H}{B}\right)\right\}\left[1 - \exp\{-a(j_w + j_f + i'_d X)\}\right] dX \tag{6.57}$$

The ground reaction P can be calculated from Eqs. (6.5), (6.6) and (6.37), (6.41) as follows:

$$P = \frac{1}{\cos \theta'_{ti}}(W \cos \beta - T_3 \sin \theta'_{ti}) \tag{6.58}$$

Following this, an optimum effective driving force T_{4opt} can be defined as the effective driving force T_4 at the optimum slip ratio i_{opt} when the effective driving energy E_4 takes on a maximum value for a constant circumferential speed V' of the flexible track belt. Likewise, a tractive power efficiency E_d can be defined as follows:

$$E_d = (1 - i_d)\frac{T_4}{T_1} \tag{6.59}$$

6.3.2   At braking state

In the flow chart of Figure 6.5, several procedures for calculating the amounts of slip sinkage s'_{fs}, s'_{rs} that occur during braking action and for determining the total amounts of sinkage of front-idler s'_{fi} and the rear sprocket s'_{ri} and the drag T_3 that develop under the flexible track belt are outlined in detail.

The amount of slip sinkage s'_{fs} at the front-idler as expressed in Eq. (6.31) can be calculated by substituting the amount of slippage j_{fs} as computed in the next equation, into Eq. (6.38).

$$j_{fs} = -(R_f + H)i_b \sin \theta_f \tag{6.60}$$

The amount of slip sinkage s'_{ri} at the rear sprocket as expressed in Eq. (6.32) can be calculated by substituting the amount of slippage j_s, as calculated in the next equation, into Eq. (6.40).

$$\begin{aligned}j_s &= -i'_b D \\ i'_b &= (1 + i_b)\cos \theta'_{ti} - 1\end{aligned} \tag{6.61}$$

As a next step, the drag T_3 that develops during braking action can be calculated from Eq. (6.41). For the situation where $j_f + j_w \geq 0$, the respective drags T_{mb} and T_{ms} acting on the base area and the side parts of the grousers of the main part of the flexible track belt can be calculated using the following equations – as shown in the previous Section 5.3.3,

$$T_{mb} = 2B \int_0^{DD} \left\{ \tau_p - k_0(j_q - j_m)^{n_0} \right\} dX$$

$$- 2B \int_{DD}^{D} \left\{ m_c + m_f \cdot p_i'(X) \right\} \left[1 - \exp\{-a(j_q - j_m)\} \right] dX \tag{6.62}$$

$$T_{ms} = 4H \int_0^{DD} \left\{ \tau_p - k_0(j_q - j_m)^{n_0} \right\} dX$$

$$- 4H \int_{DD}^{D} \left\{ m_c + m_f \frac{p_i'(X)}{\pi} \cot^{-1}\left(\frac{H}{B}\right) \right\} \left[1 - \exp\{-a(j_q - j_m)\} \right] dX \tag{6.63}$$

where

$$j_m = \frac{i_b'}{1 + i_b'} X + j_f + j_w$$

$$j_f = (R_f + H) \left[\theta_f - \frac{1}{1 + i_b} \left\{ \sin(\theta_f + \theta_{ti}') - \sin \theta_{ti}' \right\} \right] \tag{6.64}$$

and DD can be given for $j_f + j_w > 0$ as follows:

$$DD = -(j_f + j_w - j_q)\left(1 + \frac{1}{i_b'}\right) \tag{6.65}$$

Further, the drags T_{fb} and T_{fs} acting on the base area and the side parts of the grousers on the contact part of the front-idler can be calculated by substituting the following expression for the amount of slippage $j_f(\theta)$ into the previous Eqs. (5.160) and (5.161), respectively.

$$j_f(\theta) = (R_f + H) \left[(\theta_f - \theta) - \frac{1}{1 + i_b} \left\{ \sin(\theta_f + \theta_{ti}') - \sin(\theta + \theta_{ti}') \right\} \right] \tag{6.66}$$

The drags T_{rb} and T_{rs} acting on the base area and the side parts of the grousers on the contact part of the rear sprocket can be also calculated by substituting the following amount of slippage $j_r(\delta)$ into the previous Eqs. (5.169) and (5.170), respectively.

$$j_r(\delta) = (R_r + H)\left\{ (\theta_{ti}' - \delta) - \frac{1}{1 + i_b}(\sin\theta_{ti}' - \sin\delta) \right\} + \frac{i_b'D}{1 + i_b'} + j_f + j_w \tag{6.67}$$

Additionally, when the eccentricity e_i of the ground reaction falls outside the middle-third of the main contact part of the flexible track belt during braking action, the drags T_{mb}' and T_{ms}' that act on the base area and the side parts of the grousers of the main contact part of the track belt and the drags T_{rb}' and T_{rs}' that act on the base area and the side parts of the grousers of the contact part of the rear sprocket for $e_y > 1/6$, and the thrusts T_y'' and T_y'' acting on the base area and the side parts of the grousers of the main contact part of the track belt for $e_i < 1/6$ can be calculated in the same way as has been previously discussed in Chapter 5. The process is as shown in the following equations.

For $s'_{f0i} < 0 < H < s'_{r0i}$

$$T'_{mb} = -2B \int_{D-L}^{D} \{m_c + m_f p'_i(X)\}$$

$$\times \left\| 1 - \exp\left[a \left\{ j_w + \frac{i'_b}{1 + i'_b}(X - D + L) \right\} \right] \right\| dX \tag{6.68}$$

$$T'_{ms} = -4H \int_{D-L}^{D} \left\{ m_c + m_f \frac{p'_i(X)}{\pi} \cot^{-1}\left(\frac{H}{B}\right) \right\}$$

$$\times \left\| 1 - \exp\left[a \left\{ j_w + \frac{i'_b}{1 + i'_b}(X - D + L) \right\} \right] \right\| dX \tag{6.69}$$

The values of T'_{rb} and T'_{rs} can be calculated by substituting the amount of slippage $j_r(\delta)$, given in the next equation, into Eqs. (6.67) and (6.68) respectively.

$$j_r(\delta) = (R_r + H) \left\{ (\theta'_{ti} - \delta) - \frac{1}{1 + i_b}(\sin \theta'_{ti} - \sin \delta) \right\} + \frac{i'_b L}{1 + i'_b} + j_w \tag{6.70}$$

For $s'_{foi} > H > 0 > s'_{roi}$ and $j_f + j_w < 0$

$$T''_{mb} = -2B \int_{0}^{LL} \{m_c + m_f p'_i(X)\}$$

$$\times \left\| 1 - \exp\left[a \left\{ j_w + \frac{i'_b}{1 + i'_b}(X - D + L) \right\} \right] \right\| dX \tag{6.71}$$

$$T''_{ms} = -4H \int_{0}^{LL} \left\{ m_c + m_f \frac{p'_i(X)}{\pi} \cot^{-1}\left(\frac{H}{B}\right) \right\}$$

$$\times \left\| 1 - \exp\left[a \left\{ j_w + \frac{i'_b}{1 + i'_b}(X - D + L) \right\} \right] \right\| dX \tag{6.72}$$

For $j_f + j_w \geq 0$, it is necessary to calculate the values of T''_{mb} and T''_{ms} by classifying them into either of three cases i.e. $0 \leq DD \leq LL$, $LL < DD \leq D$ or $DD \geq D$. In these expressions the value of DD is as given in Eq. (6.63) – as mentioned in the Section 5.3.3.

Further, in the case where $s'_{foi} > H > 0 > s'_{roi}$, it is necessary to carefully calculate the values of T'''_{mb} given in Eq. (6.56) and T'''_{ms} given in Eq. (6.57) during driving action and, likewise, the values of T'''_{mb} given in Eq. (6.71) and T'''_{ms} given in Eq. (6.72) during braking action. The calculation may be done in the same way as shown in the previous Eqs. (5.118) $\sim$ (5.121) and Eqs. (5.149) $\sim$ (5.152). In these calculations, specific consideration must be given to the fact that the cohesive factor m_c will apply on the track belt in the range of $LL < X < D$ as a consequence of the touching of the rear sprocket following the passage of the front-idler on the terrain even if the normal contact pressure $p'_i(X)$ becomes zero.

Next, the ground reaction P can be calculated from Eq. (6.58) in the same manner as has been discussed for the driving state. The optimum effective braking force T_{4opt} during braking action can be defined as the effective braking force T_4 at the optimum skid i_{bopt} when the effective input energy $|E_1|$ takes a maximum value at a constant vehicle speed V. In addition, the braking power efficiency E_b can be defined as:

$$E_b = \frac{1}{1 + i_b} \cdot \frac{T_4}{T_1} \tag{6.73}$$

6.4   THEORY OF STEERING MOTION

Figures 6.6(a), (b), (c) give front, side-elevation and plan views of a tracked vehicle. Also shown are the primary dimension and the systems of forces that act on the vehicle when it is running on a weak flat terrain with a turning motion. The dimension D is the contact length of the track belt, B is the track width, H is the grouser height and G_p is the grouser pitch. The dimension C is the distance between the center-lines of the inner and the outer track. The parameter r_f is the radius of the front-idler. Likewise r_r is the radius of the rear sprocket. The sinkages s_{fi}, s_{fo} and s_{ri}, s_{ro} are the amounts of sinkage of the front-idler and the rear sprocket for the inner and outer tracks, respectively. W is the total vehicle weight whilst W_i and W_o are the components of the vehicle weight that are distributed to the inner and to the outer track belts, respectively.

In terms of the various forces that act on the machine, T_{4lat} is the lateral effective tractive effort and T_L is an additional lateral force acting at a point F which may be transmitted from a second connecting vehicle, depending on the direction of the effective tractive effort. From

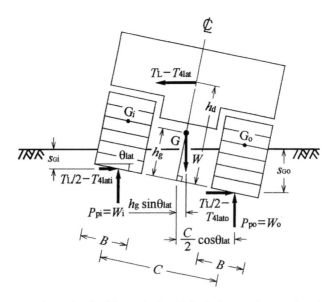

Figure 6.6(a). Front view of a flexible tracked vehicle during turning motion showing principal dimensions and forces.

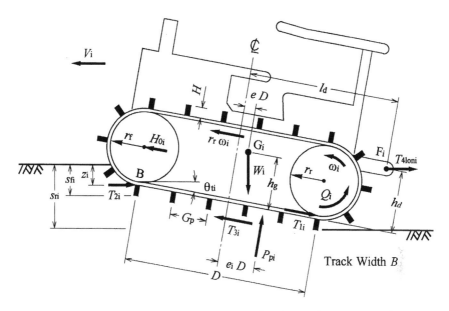

Figure 6.6(b). Side view (innertrack) of a flexible tracked vehicle during turning motion.

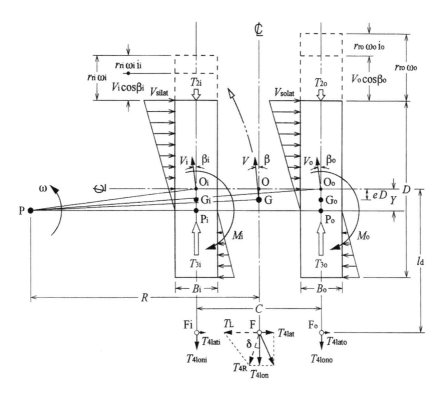

Figure 6.6(c). Plan (resultant slip velocity V_{si}, V_{so}) view of a flexible tracked vehicle during turning motion.

Figure 6.6(a), the values of W_i and W_o can be calculated by use of the following equations:

$$W_i = W \left\{ \frac{1}{2} - \frac{h_g(\tan \theta_{lat})}{C} \right\} + (T_L - T_{4lat}) \left\{ \frac{h_d}{C} + \frac{(\tan \theta_{lat})}{2} \right\} - \left(\frac{T_L}{2} - T_{4lati} \right) \tan \theta_{lat}$$

$$(6.74)$$

$$W_o = W \left\{ \frac{1}{2} - \frac{h_g(\tan \theta_{lat})}{C} \right\} - (T_L - T_{4lat}) \left(\frac{h_d}{C} - \frac{\tan \theta_{lat}}{2} \right) - \left(\frac{T_L}{2} - T_{4lato} \right) \tan \theta_{lat}$$

$$(6.75)$$

where θ_{lat} is the angle of lateral inclination of vehicle.

$$\theta_{lat} = \sin^{-1}\{(s_{G0} - s_{Gi})/C\}$$

$$s_{Gi} = s_{fi} + (s_{ri} - s_{fi})(0.5 + e) \qquad (6.76)$$

$$s_{G0} = s_{f0} + (s_{r0} - s_{f0})(0.5 + e)$$

Here, h_g is the height of the center of gravity G of the vehicle measured from the bottom track belt, h_d is the height of the application point of the resultant effective tractive effort T_{4R}, and e is the eccentricity of the center of gravity G.

The factor eD is the amount of eccentricity measured from the center line of the vehicle. Also, W_i and W_o can be assumed to act on the same position G_i, G_o for inner and outer track as the position G.

As shown in Figure 6.6(b), the forces T_{1i} and T_{1o} are the driving forces transmitted from the torques Q_i and Q_o of the rear sprocket, which are applied on the track belt at the bottom-dead-center of the rear sprocket for the inner and the outer track belts, respectively. T_{2i} and T_{2o} are the locomotion resistances i.e. the compaction resistances acting in front of the inner and outer track belts at the depths z_i and z_o, which can be calculated from each rut depth, s_{ri} and s_{ro} [2].

The forces T_{3i} and T_{3o} are the thrusts developed along the track belt under the interface between the terrain and the grousers of the inner and outer track belts. These can be calculated as the integral of the shear resistance of the soil. Usually, the driving force $T_{1i(o)}$ can be assumed to be the same as the thrust $T_{3i(o)}$, which depends on the shearing characteristics of the terrain.

T_{4loni} and T_{4lono} are the effective tractive forces acting on the inner and outer tracks. These can be calculated from a force balance [3] as shown in the following expression:

$$T_{4loni(o)} = \frac{T_{3i(o)}}{\cos \theta_{ti(o)}} - W_{i(o)} \tan \theta_{ti(o)} - T_{2i(o)} \qquad (6.77)$$

where $\theta_{ti(o)}$ is the angle of longitudinal inclination of the inner and the outer track.

The longitudinal effective tractive effort T_{4lon} acting in the longitudinal direction of the vehicle can be expressed as the sum of each effective tractive force T_{4loni} and T_{4lono}, as follows:

$$T_{4lon} = T_{4loni} + T_{4lono} \qquad (6.78)$$

The lateral effective tractive effort T_{4lat} acting in the lateral direction of the vehicle is given as the integral of the shear resistance $\tau_{i(o)lat}(X)$ that develops along the inner and outer track belts, as follows:

$$T_{4lat} = T_{4lati} + T_{4lato}$$

$$= B \int_0^D \{\tau_{ilat}(X) + \tau_{olat}(X)\} \, dX \tag{6.79}$$

where X is the distance from the front part of the track belt. Theoretically, T_{4lat} should take a value of zero because the turning moment resistance is only developed by the difference of the effective tractive effort of inner and outer track, but it has some negligibly small value due to experimental errors in determining the terrain-track system constants in the lateral direction of the track plate. In these calculations, the centrifugal force is not considered because of its negligibly small value at low vehicle speeds. The angle δ of the resultant effective tractive effort T_4 is determined as:

$$\delta = \tan^{-1}\left\{ (T_L - T_{4lat})/T_{4lon} \right\} \tag{6.80}$$

The reactions P_{pi} and P_{po} are the resultant normal ground reaction forces applied on the inner and the outer track, to which the amounts of eccentricity are e_iD and e_oD, respectively. H_{0i} and H_{0o} are the initial track tensions for the inner and the outer track belts.

As depicted in Figure 6.6(c), F is the point of application of the resultant effective tractive effort T_4 on the vehicle. This is composed of a longitudinal component T_{4lon}, and a lateral one T_{4lat}. There is also an additional lateral force T_L, for which the height h_d is the distance measured from the bottom-dead-center of the rear sprocket and l_d is the distance from the center line of the vehicle. Additionally, T_{4loni}, T_{4lati} and T_{4lono}, T_{4lato} can be assumed to act at the same position F_i and F_o for the inner and outer tracks as the position F on the longitudinal plane.

The point P is the turning center of the tracked vehicle, measured from the center point O of the tracked vehicle, i.e. R is the turning radius of the vehicle and Y is the deviation from the lateral center line of the vehicle. The elements M_i, M_o are the turning resistance moments acting around the point P_i and P_o of the inner and outer track. The deviations of these, from the center point O_i and O_o are $Y_i = e_iD$ and $Y_o = e_oD$ for the inner and outer tracks, respectively.

The longitudinal effective tractive effort T_{4ion} can be also derived from the following moment balance equation:

$$RT_{4lon} + (l_d - Y)(T_L - T_{4lat})$$

$$= \left(R - \frac{C}{2} \right) \left(\frac{T_{3i}}{\cos \theta_{ti}} - W_i \tan \theta_{ti} - T_{2i} \right)$$

$$+ \left(R + \frac{C}{2} \right) \left(\frac{T_{3o}}{\cos \theta_{to}} - W_o \tan \theta_{to} - T_{2o} \right) - M_i - M_o$$

$$= \left(R - \frac{C}{2} \right) T_{4loni} + \left(R + \frac{C}{2} \right) T_{4lono} - M_i - M_o \tag{6.81}$$

$$Y = (e_1 + e_2) D/2 \tag{6.82}$$

Substituting Eq. (6.78) into the above equation, we get:

$$T_{4lono} - T_{4loni} = 2\{M_i + M_0 + (l_d - Y)(T_L - T_{4lat})\}/C \tag{6.83}$$

When δ is zero, the difference between T_{4lono} and T_{4loni} can be calculated as $2(M_i + M_o)/C$ using Eq. (6.83) at $T_L = T_{4lat} \approx 0$. When the additional lateral force T_L becomes large, the difference between T_{4lono} and T_{4loni} should be increased.

The resultant effective tractive effort T_{4R} can be calculated as follows:

$$T_{4R} = \{(T_{4lon})^2 + (T_L - T_{4lat})^2\}^{1/2} \tag{6.84}$$

The parameter β is the slip angle of the center of gravity of the vehicle and β_i and β_o are the slip angles of the inner and outer tracks, respectively. The value of these parameter is given by:

$$\beta_i = \tan^{-1}\{Y/(R - C/2)\} \tag{6.85}$$

$$\beta_0 = \tan^{-1}\{Y/(R + C/2)\} \tag{6.86}$$

$$\beta = \tan^{-1}\{(Y - eD)/R\} \tag{6.87}$$

The compaction resistance $T_{2i(o)}$ can be computed from the following expression:

$$T_{2i(o)} = B \int_0^{Sri(o)} k_1 s^{n_1} ds \tag{6.88}$$

6.4.1   Thrust and steering ratio

For the inner and outer tracks, the longitudinal slip velocity $V_{si(o)lon}$ and the longitudinal amount of slippage $j_{i(o)ion}(X)$ at a distance X from the front part of the inner and outer tracks can be calculated through use of the following two equations:

$$V_{si(o)lon} = r_{ri(o)}\omega_{i(o)} - V_{i(o)} \tag{6.89}$$

$$j_{i(o)lon}(X) = \int_0^t (r_{ri(o)}\omega_{i(o)} - V_{i(o)})\,dt = i_{i(o)}X \tag{6.90}$$

where t is the movement-time of the track belt i.e. $X/r_{ri(o)\omega i(o)}$ from the front part to a point X, and $i_{i(o)}$ is the slip ratio of the inner and outer tracks as will be discussed later.

The longitudinal shear resistance of the soil that develops under the inner and outer track belts $\tau_{i(o)lon}(X)$ at points N_i and N_o may be calculated as:

$$\tau_{i(o)lon}(X) = (m_{clon} + p_{i(o)}(X)m_{flon})[1 - \exp\{-a_{lon}j_{i(o)lon}(X)\}] \tag{6.91}$$

Thence, the main thrust of the inner and outer tracks $T_{3i(o)}$ can be calculated as:

$$T_{3i(o)} = B \int_0^D \tau_{i(o)lon}(X)\,dX \tag{6.92}$$

When the circumferential speed of the rear sprocket of inner track and outer track is set to be $r_{ri}\omega_i$ and $r_{ro}\omega_o$, a steering ratio ε may be defined as follows:

$$\varepsilon = r_{ro}\omega_o/r_{ri}\omega_i \tag{6.93}$$

Again, another steering ratio ε' can be defined as in the following equation, where the speed of the inner and the outer tracks are set to be V_i and V_o at slip angles β_i and β_o, respectively:

$$\varepsilon' = V_o \cos \beta_o / V_i \cos \beta_i \qquad (6.94)$$

In this case, the slip ratios of the inner and outer tracks i_i, i_o are expressed as:

$$i_i = 1 - V_i \cos \beta_i / r_{ri} \omega_i \qquad (6.95)$$

$$i_o = 1 - V_o \cos \beta_o / r_{ro} \omega_o \qquad (6.96)$$

when both the track belts are in the driving state.

Substituting the above equations into Eq. (6.93), the following relationship can be derived.

$$\varepsilon' = \varepsilon (1 - i_o)/(1 - i_i) \qquad (6.97)$$

The turning speed of the tracked vehicle V at the center of gravity G and the running speed of the inner and outer tracks V_i and V_o may be calculated as follows:

$$V = \omega \sqrt{R^2 + (Y - eD)^2} \qquad (6.98)$$

$$V_i = r_{ri} \omega_i (1 - i_i)/\cos \beta_i = \omega \sqrt{(R - C/2)^2 + Y^2} \qquad (6.99)$$

$$V_o = r_{ro} \omega_o (1 - i_o)/\cos \beta_o = \omega \sqrt{(R + C/2)^2 + Y^2} \qquad (6.100)$$

$$\omega = r_{ri} \omega_i (1 - i_i)/(R - C/2) = r_{ro} \omega_o (1 - i_o)/(R + C/2) \qquad (6.101)$$

where ω is the steering angular velocity of the tracked vehicle.

Eliminating the steering angular velocity ω, the turning radius of the tracked vehicle R can be determined as:

$$R = \frac{C \left\{ r_{ro} \omega_o (1 - i_o) + r_{ri} \omega_i (1 - i_i) \right\}}{2 \left\{ r_{ro} \omega_o (1 - i_o) - r_{ri} \omega_i (1 - i_i) \right\}} \qquad (6.102)$$

6.4.2   Amount of slippage in turning motion

Calculation of the lateral slip velocity between a track and a soil and the amount of lateral slippage of soil under a track belt in turning motion is required to determine the turning resistance moment of the inner and outer tracks.

Figure 6.7 shows the resultant slip velocity $V_{si(o)}$ whose components are $r_{ri(o)} \omega_{i(o)}$ in the longitudinal direction and $\omega \overline{PN_i}$ and $\omega \overline{PN_o}$ in the tangential direction. As a consequence, the lateral slip velocity $V_{si(o)lat}(X)$ of the inner and outer tracks at arbitrary points N_i and N_o may be given, cf. also Figure 6.6(c), as the lateral component of the resultant slip velocity.

$$\begin{aligned} V_{si(o)lat}(X) &= \omega \sin \alpha \sqrt{(R \pm C/2)^2 + (D/2 - X + Y)^2} \\ &= \omega(D/2 - X + Y) \\ &= V_{i(o)} \frac{D/2 - X + Y}{\sqrt{(R \pm C/2)^2 + Y^2}} \end{aligned} \qquad (6.103)$$

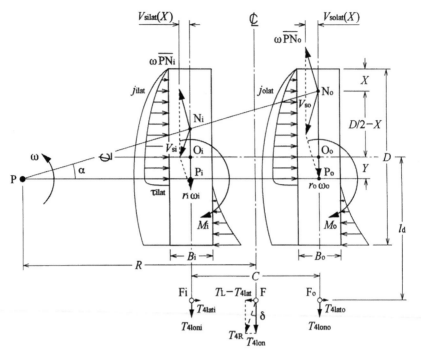

Figure 6.7. Lateral amount of slippage j_{ilat}, j_{olat} and shear resistance τ_{ilat}, τ_{olat} for inner and outer track.

Thence, the lateral amount of slippage $j_{i(o)lat}(X)$ can be calculated as:

$$j_{i(o)lat}(X) = \int_0^t V_{si(o)lat}(X)\,dt$$

$$= \frac{V_{i(o)}}{\sqrt{(R \pm C/2)^2 + Y^2}} \int_o^t (D/2 - X + Y)\,dt$$

$$= \frac{V_{i(o)}}{r_{ri(o)}\omega_{i(o)}\sqrt{(R \pm C/2)^2 + Y^2}} \int_o^X (D/2 - X + Y)\,dX$$

$$= \frac{1 - i_{i(o)}}{\cos\beta_{i(o)}\sqrt{(R \pm C/2)^2 + Y^2}}(D/2 + Y - X/2)X$$

$$= \frac{1 - i_{i(o)}}{R \pm C/2}(D/2 + Y - X/2)X \qquad (6.104)$$

where t is the time of travel of a grouser from the front part of the track belt to arbitrary points N_i and N_o under the inner and outer track. This time is given as $X/r_{ri(o)}\omega_{i(o)}$. As a consequence, it is evident that the lateral amount of slippage $j_{i(o)lat}(X)$ takes on a value of zero at $X = 0$ and $D + 2Y$. It also takes a maximum value at $X = D/2 + Y$ for both inner and outer tracks.

6.4.3   Turning resistance moment

The lateral shear resistance $\tau_{i(o)lat}(X)$ that develops along a track belt at points N_i and N_o, under an inner and outer track, can be calculated through use of the following expression.

$$dj_{i(o)lat}(X)/dX \geq 0$$

$$\tau_{i(o)lat}(X) = \{m_{clat} + p_{i(o)}(X)m_{flat}\}\left[1 - \exp\{-a_{lat}j_{i(o)lat}(X)\}\right] \tag{6.105}$$

$$dj_{i(o)lat}(X)/dX < 0 \quad j_q \leq j_{i(o)lat}(X) \leq j_p$$

$$\tau_{i(o)lat}(X) = \tau_p - k_3(j_p - j)^{n_3} \tag{6.106}$$

$$dj_{i(o)lat}(X)/dX < 0 \quad j_{i(o)lat}(X) < j_q$$

$$\tau_{i(o)lat}(X) = -\{m'_{clat} + m_{flat}\,'p_{i(o)}(X)\}\left[1 - \exp\{-a'_{lat}\{j_q - j_{i(o)lat}(X)\}\}\right] \tag{6.107}$$

where $j_p = j_{i(o)lat}(D/2 + Y)$, $t_p = t_{i(o)lat}(D/2 + Y)$, $j_q = j_p - (t_p/k_3)^{1/n_3}$, and $p_{i(o)}(X)$ is the normal pressure distribution under the inner and outer track belt.

The parameters m_{clat}, m_{flat} and a_{lat}, k_3, n_3, and m'_{clat}, m'_{flat} and a'_{lat} are the lateral terrain-track system constants which were measured – including the bulldozing resistance of the track model plate – via the track plate traction test. It is noted here that the reaction of the additional lateral force T_L is automatically included in this distribution of lateral shear resistance.

Following this, the turning resistance moments M_i and M_o that are exerted around the turning points P_i and P_o of the inner and outer track, can be calculated – including the amount of sinkage of the track plate – as follows:

$$M_{i(o)} = B \int_0^D \tau_{i(o)lat}(X)(D/2 - X + Y)\,dX \tag{6.108}$$

The total turning resistance moment M is given by the following equation:

$$M = M_i + M_0 \tag{6.109}$$

The energy equilibrium equation for the straight forward motion of a tracked vehicle has already been presented [4]. For machines in turning motion, the input energy $E_{1i(o)}$ supplied by the driving torque can be equated to the sum of: the compaction energy $E_{2i(o)}$ required to make a rut under the track belt, the slippage energy $E_{3i(o)}$ required to develop a thrust along the bottom of the track belt, the effective tractive effort energy $E_{4i(o)}$, and the turning moment energy $E_{5i(o)}$ for the inner track and the outer track. That is,

$$E_{1i(o)} = E_{2i(o)} + E_{3i(o)} + E_{4i(o)} + E_{5i(o)} \tag{6.110}$$

where

$$E_{1i(o)} = T_{1i(o)}V_{i(o)}\cos\beta_{i(o)}/(1 - i_{i(o)})$$

$$E_{2i(o)} = T_{2i(o)}V_{i(o)}\cos\beta_{i(o)}$$

$$E_{3i(o)} = T_{3i(o)}\{1/(1 - i_{i(0)}) - 1/(\cos\theta_{ti(o)})\} V_{i(o)} \cos\beta_{i(o)}$$
$$+ W_{i(0)} V_{i(o)} \cos\beta_{i(o)} \tan\theta_{ti(o)}$$

$$E_{4i(o)} = T_{4loni(o)} V_{i(o)} \cos\beta_{i(o)} + \{T_L/2 - T_{4lati(o)}\} (l_d - e_{i(o)}D)\omega$$

$$E_{5i(o)} = \omega M_{i(o)}$$

Then the total input energy E_1 of the vehicle and the total output energy of the compaction energy E_2, the slippage energy E_3, the effective tractive effort energy E_4 and the turning moment energy E_5 of the vehicle can be given as in the equations:

$$E_1 = E_2 + E_3 + E_4 + E_5 \tag{6.111}$$

$$E_1 = E_{1i} + E_{1o} \tag{6.112}$$

$$E_2 = E_{2i} + E_{2o} \tag{6.113}$$

$$E_3 = E_{3i} + E_{3o} \tag{6.114}$$

$$E_4 = E_{4i} + E_{4o} \tag{6.115}$$

$$E_5 = E_{5i} + E_{5o} \tag{6.116}$$

This energy equilibrium equation can also be proved theoretically by using the force and moment balance equations as mentioned above. An optimum resultant effective tractive effort T_{4Ropt} may be defined as the resultant effective tractive effort at the optimum combination of slip ratio i_{iopt} of the inner track and i_{oopt} of the outer track which takes the maximum value of the effective tractive effort energy E_{4max}. A tractive power efficiency E_d may be developed as follows:

$$E_d = E_4/E_1 \tag{6.117}$$

6.4.4   Flow chart

As illustrated in Figure 6.8, the necessary input information for a simulation analysis of a flexibly tracked machine operating on a terrain includes the weight W, the contact length D of the inner and the outer tracks, the track width B of the inner and the outer tracks, the central distance C between the inner and the outer tracks, the radius r_f of the front-idler, the radius r_r of the rear sprocket, the radius of the track roller r_m, the number of track rollers N, the grouser height H, the eccentricity of the center of gravity e of the vehicle, the height h_g of the center of gravity of the vehicle, the distance l_d between the central axis of the vehicle and the application point, the height h_d of the application point of the total effective tractive effort, the initial track belt tension H_0. Following the input of this primary geometrical and mass data, as a next step the terrain-track system constants k_1, k_2 and n_1, n_2 from the plate loading and unloading test, m_{clon}, m_{clat}; m_{flon}, m_{flat}; a_{lon}, a_{lat} from the plate traction test, and c_{0lon}, c_{0lat}; c_{1lon}, c_{1lat}; c_{2lon}, c_{2lat} from plate slip sinkage test need to be read in as data.

At rest, the static amount of sinkage $s_{fi} = s_{fo}$ and $s_{ri} = s_{ro}$, the linear distribution of static amount of sinkage $s_i(X)$ of inner track which equals $s_o(X)$ of outer track at the distance X from the contact point of front-idler on the main part of track belt, the angle of longitudinal inclination of inner and outer track $\theta_{ti} = \theta_{to}$, the contact pressure $p_{fi} = p_{fo}$ at the front-idler and $p_{ri} = p_{ro}$ at the rear sprocket, the nonlinear normal pressure distribution $p_i(X)$ of

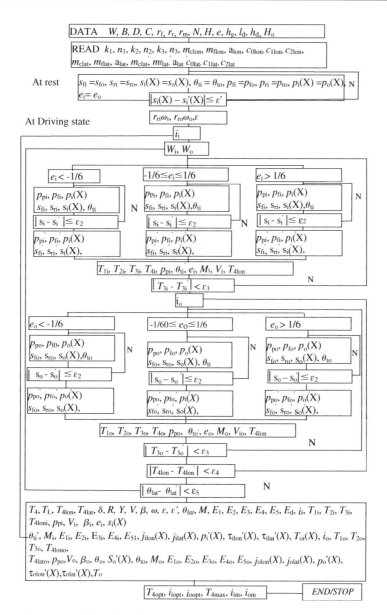

Figure 6.8. Flow chart of flexible track.

the inner track which equals $p_o(X)$ of the outer track, and the eccentricity $e_i = e_o$ of the resultant force $P_{pi} = P_{po}$ for the assumed rigid track belt can be iteratively calculated until the distribution of static amount of sinkage $s_i(X) = s_o(X)$ is determined.

For a given angular velocity of the rear sprocket ω_i for the inner track and ω_o for the outer track, and for a steering ratio ε, both the tractive performances of the inner and the outer track during driving action can be calculated for each combination of slip ratio i_i and i_o.

First of all, the tractive performance of the inner track is calculated for the slip ratio i_i of the inner track, assuming that the distributed vehicle weight W_i and W_o equals half the vehicle's weight i.e. $W/2$ respectively.

For the above calculated eccentricity e_i, the resultant normal force P_{pi}, the contact pressure p_{fi}, p_{ri} and $p_i(X)$, the total amount of sinkage s_{fi}, s_{ri} and $s_i(X)$ including the amount of slip sinkage, and the angle θ_{ti} can be iteratively calculated (depending on the three kinds of procedure calculations streams which are divided according to the value of eccentricity e_i) until the distribution of the total amount of sinkage $s_i(X)$ is determined. In order to transform the above results that have been calculated for an assumed rigid track belt into those of an actual flexible track belt, the normal contact pressure distribution and the distribution of the total amount of sinkage need to be changed to p'_{fi}, p'_{ri}, and $p'_i(X)$ and s'_{fi}, s'_{ri}, and $s'_i(X)$ for the flexible track belt considering the initial track belt tension H_{0i}. This process has been discussed in a previous paper [5]. Then, the driving force T_{1i}, the compaction resistance T_{2i}, the thrust T_{3i} from Eq. (6.92), the effective tractive force T_{4i} from Eq. (6.77), the angle θ_{ti}, the eccentricity e_i of the normal resultant force P_{pi}, the turning resistance moment M_i from Eq. (6.108), the running speed V_i from Eq. (6.99), and the longitudinal effective tractive effort of the vehicle T_{4lon} from Eq. (6.78) can be iteratively calculated until the thrust T_{3i} is determined.

As a next step, the tractive performance of the outer track is calculated for the slip ratio i_o of the outer track assuming that the distributed vehicle weight W_o equals $W/2$. For the above calculated eccentricity $e_o = e_i$, the ground reaction P_{po}, the contact pressure p_{fo}, p_{ro} and $p_o(X)$, the total amount of sinkage s_{fo}, s_{ro} and $s_o(X)$ including the amount of slip sinkage, and the angle θ_{to} can be repeatedly calculated (depending on the three kinds of procedure calculations streams which are divided according to the value of eccentricity e_i) until the distribution of the total amount of sinkage $s_o(X)$ is determined. In order to transform the above results calculated for the assumed rigid track belt to the actual flexible track belt, the normal contact pressure distribution and the distribution of the total amount of sinkage should be changed to p'_{fo}, p'_{ro}, and $p'_o(X)$ and s'_{fo}, s'_{ro}, and $s'_o(X)$ for the flexible track belt considering the initial track belt tension H_{0o}. Then, the driving force T_{1o}, the compaction resistance T_{2o}, the thrust T_{3o} from Eq. (6.92), the effective tractive force T_{4o} from Eq. (6.77), the angle θ_{to}, the eccentricity e_o of the ground reaction P_{po}, the turning resistance moment M_o from Eq. (6.108), the running speed V_o from Eq. (6.100), and the longitudinal effective tractive effort of the vehicle T_{4lon} from Eq. (6.78) can be iteratively calculated until the thrust T_{3o} is determined.

Thereafter, the actual slip ratio i_o of the outer track for the given slip ratio i_i of the inner track can be calculated repeatedly by using the two division method until the longitudinal effective tractive effort of the vehicle T_{4lon} from Eq. (6.83) is determined. After that, the real distributions of vehicle weight to the inner and the outer track W_i and W_o can be calculated recursively from Eqs. (6.74) and (6.75) until the real angle of lateral inclination of the vehicle θ_{lat} is determined.

Following this, the tractive performance of the vehicle i.e. the resultant effective tractive effort T_{4R} from Eq. (6.84) which is composed of the longitudinal effective tractive effort T_{4lon} from Eq. (6.78) and the lateral effective tractive effort T_{4lat} from Eq. (6.79), and the angle δ of the vehicle, the position of the turning center of the turning radius of the vehicle R from Eq. (6.102) and Y from Eq. (6.82), the running speed V and the angle β, the steering ratio ε from Eq. (6.93) and ε' from Eq. (6.94), the angle of lateral inclination of the vehicle

θ_{lat}, the total turning resistance moment M from Eq. (6.109), and the total amount of the input energy E_1, the compaction energy E_2, the slippage energy E_3, the effective tractive effort energy E_4 and the turning moment energy E_5 from Eqs. (6.112)~(6.116), and the tractive power efficiency E_d from Eq. (6.117) can be determined for each combination of slip ratio i_i and i_o.

In addition, the tractive performances of the inner track and the outer track i.e. the slip ratio $i_{i(o)}$, the driving force $T_{1i(o)}$, the compaction resistance $T_{2i(o)}$, the thrust $T_{3i(o)}$, the effective tractive force $T_{4i(o)}$, the running speed $V_{i(o)}$ and the angle $\beta_{i(o)}$, the eccentricity $e_{i(o)}$ of the resultant normal force $P_{pi(o)}$, the distribution of the total amount of sinkage $s_{i(o)}(X)$, the angle of longitudinal inclination $\theta_{ti(o)}$, the turning resistance moment $M_{i(o)}$, the distribution of the longitudinal amount of slippage $j_{i(o)lon}(X)$ and the lateral amount of slippage $j_{i(o)lat}(X)$, the normal contact pressure distribution $p_{i(o)}(X)$, the shear resistance distribution $\tau_{i(o)}(X)$, the distribution of track tension $T_{0i(o)}(X)$, the input energy $E_{1i(o)}$, the compaction energy $E_{2i(o)}$, the slippage energy $E_{3i(o)}$, the effective tractive effort energy $E_{4i(o)}$ can be determined in detail. Finally, the optimum effective tractive effort T_{4Ropt} and the optimum combination of slip ratio i_{iopt} of the inner track and i_{oopt} of the outer track, can be determined from Eq. (6.117).

6.5   SOME EXPERIMENTAL STUDY RESULTS

In general, the soft terrain under a flexible track belt (which is the typical undercarriage system on many pieces of construction machinery such as bulldozer or tractors) is taken to failure or is transformed into the plastic state. This occurs, because the contact pressure of the track plate reaches several times that of the average contact pressure and exceeds the bearing capacity of the terrain when the axle load of a road roller is supported by only a few track plates located just under the road roller. It is generally expected that the terrain under the flexible track belt moves into a state of active earth pressure when the amount of deflection of track belt becomes convex to the terrain, while it moves into a state of passive earth pressure for a concave deflection of track belt to the terrain [6]. In this section, various traffic performance aspects of a flexible tracked vehicle, especially the aspect of characteristic contact pressure distribution under the flexible track belt during self-propelling action and whilst operating under traction, are presented.

6.5.1   During self-propelling operation

From the literature, it is well known that the contact pressure distribution under a flexible track belt has a wavy sinusoidal distribution having peak values under the individual road rollers.

Rowland [7] reported that the maximum contact pressure measured at a depth of 23 cm was 1.2 to 2.0 times larger than the average contact pressure of a flexible tracked machine equipped with several types of undercarriage, road rollers and suspension systems during self-propelling states on loose accumulated sandy terrain and soft terrain composed of clayey soil or muskeg etc. Cleare [8] measured the maximum contact pressure at depth of 30 cm of a soft silty terrain as being 1.3 times larger than the average contact pressure of a flexible tracked vehicle. Again, Fujii et al. [9] measured the distribution of contact pressure

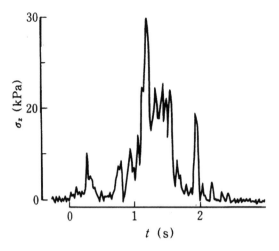

Figure 6.9. Relationship between contact pressure σ_z under flexible track belt at depth of 14.5 cm and passive time t (self propelling state) [5].

under the flexible track belt of a tractor of weight 113 kN, contact length of track belt 220 cm and track width 40 cm when the tractor was self-propelling on a convex sandy terrain of bulk density 16.3 kN/m³ and water content $19 \sim 24\%$. Figure 6.9 shows an example of contact pressure distribution under the flexible track belt. The pressures were measured by use of an earth pressure cell buried at a depth of 14.5 cm. In this case, the maximum contact pressure dropped down to about 0.63 times that of the average contact pressure of 64 kPa – possibly because the measured point might have deviated from the center line of the running line of the tractor.

Sofiyan et al. [10] measured directly the distribution of contact pressure under the flexible track belt of a tractor running on a sandy terrain by use of strain gauges attached directly to track links. They reported that the maximum contact pressure reached $3.0 \sim 3.5$ times larger that the average contact pressure. Thus, it becomes clear that the maximum value, period or wavy pattern of the contact pressure depends on the soil properties and the surface roughness of the terrain, the size or number of road rollers, the mechanism of connection and the suspension system, the track tension between the pin joints of the track links, the shape of track plate, the structure of track belt and so on.

There are not many available experimental studies measuring the behaviour of soil particles under a flexible track belt, but Yong et al. [11] measured the elliptical trajectories of soil particles under a rigid wheel during driving action at various slip ratios. They suggested that the relative amounts of movement between a flexible track belt and a terrain become very large and the soil particles under the flexible track belt move very significantly in the horizontal and vertical direction especially immediately under the road rollers. Experimentally, it was observed that the soil particles under a flexible track belt during self-propelling state move forward in front of the road rollers and move backward to their rear sides due to the comparatively small track tension [12]. The shear resistance acting on the flexible track belt takes on a positive or negative value in correspondence to the alternative relative amounts of slippage between the track plate and the terrain under the road rollers.

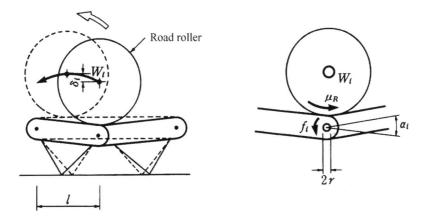

Figure 6.10. Mechanism of rolling frictional resistance of road roller on flexible track belt.

Next, some information on the rolling frictional resistance T_r of a road roller rotating on the track link of a flexible track belt will be presented. Figure 6.10 shows a mechanism illustrating the dynamic rolling movement of each road roller of axle load W_i rotating on two track links separated at the distance of one track plate length l [13]. If one assumes that the track plates, with grousers mounted on the track links, are operating on a hard terrain, each track link will rotate mutually around the link pin of radius r for a central angle of α_i during one pass of the road roller. For an amount of vertical movement δ_i of the axle of road roller during one pass of the track plate, the increment of potential energy of the road roller equates to $W_i \delta_i$. Then, an energy equilibrium equation for one pass of the road roller on the track link of length of l can be established as:

$$T_r l = \mu_R l \sum_{i=1}^{n} W_i + \sum_{i=1}^{n} W_i \delta_i + 2r \sum_{i=1}^{m} \alpha_i f_i$$

where T_r is the rolling frictional resistance, μ_R is the coefficient of rolling resistance, f_i is the frictional force between pin and bush of track link and m and n are, respectively, the number of links and track plates for one contact length of track belt.

Thence, the rolling frictional resistance of road roller on a flexible track belt T_r can be worked out as follows:

$$T_r = \mu_R \sum_{i=1}^{n} W_i + \frac{1}{l} \left(\sum_{i=1}^{n} W_i \delta_i + 2r \sum_{i=1}^{m} \alpha_i f_i \right) \tag{6.118}$$

Further, this rolling frictional resistance T_r increases the driving torque of the rear sprocket as an internal resistance, especially for the self-propelling state of vehicle when not so much track belt tension develops.

Bekker [14, 15] investigated the relations that exist between the amount of sinkage of a flexible track belt and the number of road rollers. He found that the amount of sinkage was decreased 34% by doubling the number of road rollers and that the amount of sinkage decreased rapidly when the number of rollers increased from 2 to 5. But it then tended to a constant value as the number increased from 5 to 9.

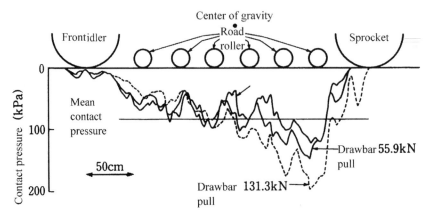

Figure 6.11. Contact pressure distribution of bulldozer under traction [19].

6.5.2   During tractive operations

The distribution of contact pressure under a flexible track belt (as installed on a typical bull-dozer or tractor under tractive operations) changes from a uniform wavy distribution to a trap-ezoidal or triangular wavy one with increases in effective tractive effort. The re-distribution of the contact pressure is caused by the occurrence of a moment around the center of gravity of the vehicle since the effective tractive effort is applied at some height to the rear part of the bulldozer. Wills [16] and Torii [17] have confirmed that the contact pressure distribution of a tractor under traction shows some wavy triangular patterns towards to the rear end of the track belt. Wong [18] observed that the distribution of the contact pressure acting on flexible track belt under traction at various slip ratios shows some wavy distribution with several peak values just under the road rollers. He also reported that the ratio of maximum contact pressure to the average one takes on a value of 9.8 for sandy terrain, 2.9 for muskeg terrain and 8.3 for snow covered terrain. Ito et al. [19] measured the contact pressure dis-tribution under a bulldozer of weight 349 kN running on a Kanto loam terrain of water content 108.7%, bulk density 13.4 kN/m^3 and cone index 1068 kPa.

Figure 6.11 shows the contact pressure distribution in self-propelled action and under driving action developing two kinds of effective tractive effort of 55.9 kN and 131.3 kN i.e. the contact pressure at the rear end of track belt in the wavy triangular distribution and the eccentricity of ground reaction tended to increase with increase in the effective tractive effort.

Also, Bekker [20] investigated the variation in the contact pressure distribution acting under a flexible track belt for a tractor of weight 30 kN running at various vehicle speeds and various degrees of effective tractive effort on a weak terrain having various water contents.

He determined that the maximum contact pressure occurred at the rear end of the flexible track belt and that it increased with increasing tractive effort and with a decrease in the water content of the terrain.

6.6   ANALYTICAL EXAMPLE

In the published literature, there have been a number of methods of theoretical analysis presented that relate to the problem of predicting the tractive performances of flexible

Table 6.1. Dimensions of bulldozer running on soft terrain.

Vehicle weight	W (kN)	50
Track width	B (cm)	100
Contact length	D (cm)	320
Average contact pressure	p_m (kPa)	7.81
Radius of front-idler	R_f (cm)	50
Radius of rear sprocket	R_t (cm)	50
Grouser height	H (cm)	12
Grouser pitch	G_p (cm)	36
Interval of road roller	R_p (cm)	40
Radius of road roller	R_m (cm)	8
Eccentricity of center of gravity of vehicle	e	-0.02
Height of center of gravity of vehicle	h_g (cm)	100
Distance of application point of effective driving or braking	l_b (cm)	300
force from central axis of vehicle	l_d (cm)	300
Height of application point of effective driving	h_b (cm)	50
or braking force	h_d (cm)	50
Initial track belt tension	H_o (kN)	9.8
Circumferential speed of track belt (during driving state)	V' (cm/s)	100
Vehicle speed (during braking state)	V (cm/s)	100

tracked vehicle on a soft terrain. For example, we have Yong's energy [21] and FEM [22] analyses concerning the interaction between the structure of a track belt and a terrain.

Likewise we have Garber's numerical analysis [23] which considers the strength of the ground, the track belt tension and the distribution of contact pressure, Wong's simulation analysis [24] and Oida's study [25] which may be used to calculate theoretically the thrust of a flexible track belt from the shear deformation properties of soil on an unit track plate.

In this section, several analytical examples relating to the trafficability of several flexible tracked vehicles running during driving and braking action on a silty loam terrain, a decomposed weathered granite sandy terrain and a snow covered terrain with considerations of the size effect of the track model plate and the initial track belt tension will be presented.

6.6.1 Silty loam terrain

(1) *Trafficability of a bulldozer running on soft terrain*
The complex tractive performance during driving action and the complex braking performance during braking action of a flexible tracked vehicle of bulldozer running on a soft terrain are developed here through use of a rigorous mathematical simulation method. The specifications of the vehicle that will be used in these studies are given in Table 6.1.

The structure of the track belt is that of a flexible rubber track belt equipped with equilateral trapezoidal grousers of trim angle $\alpha = \pi/6$ rad, contact length $L = 4$ cm and grouser pitch of $G_p = 36$ cm as shown in the previous Figure 4.16. The vehicle is specified to be running on a flat silty loam terrain of slope angle of $\beta = 0$ rad. To allow for the size effect of track model plate, the terrain-track system constants as shown in Table 6.2 have already been modified by substituting a size ratio of $N = 8$ into the previous Eqs. (4.36) ~ (4.45).

(a) In the driving state
By substituting the above mentioned vehicle specifications and terrain-track system constants as input data into the flow chart of the previous Figure 6.5, the variations of driving

Table 6.2. Terrain-track system constants for bulldozer running on soft terrain.

Track plate loading and unloading test
$p \leq 17.8$ (kPa)	$k_1 = 1.254$	$n_1 = 1.068$
$p > 17.8$ (kPa)	$k_2 = 5.513$	$n_2 = 0.895$
	$k_3 = k_4 = 30.04$	$n_3 = n_4 = 0.632$

Track plate traction test
$m_c = 3.362$ (kPa) $m_f = 0.311$ $a = 0.078$ (l/cm)

Track plate slip sinkage test
$c_0 = 0.692$ $c_1 = 0.584$ $c_2 = 0.478$

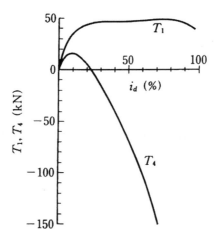

Figure 6.12. Relationship between driving force T_1 effective driving force T_4 and slip ratio i_d (during driving action).

force T_1, the effective driving force T_4, the amount of sinkage of the front-idler s'_{fi} and of the rear sprocket s'_{ri}, the angle of inclination of the vehicle θ'_{ti}, the eccentricity of the ground reaction e_i, each energy component $E_1 \sim E_4$, and the tractive power efficiency E_d, with the slip ratio i_d during driving action can be calculated.

In looking at the results, Figure 6.12 is a plot of the obtained relations between T_1, T_4 and i_d. The driving force T_1 increases initially and then tends to a constant value with increasing values of i_d. In contrast the effective driving force T_4 moves quite rapidly to a maximum value of $T_{4max} = 16.3$ kN at a slip of $i_d = 10\%$. After that it decreases gradually to take a value of zero at $i_d = 23\%$. The vehicle can not develop any traction force at a slip ratio of i_d greater than 23%.

Figure 6.13 plots the relations between the sinkages s'_{fi}, s'_{ri} and the slip ratio i_d. From the diagram it is evident that s'_{ri} increases with increasing values of i_d accompanied by increasing amounts of slip sinkage. As a consequence, the angle of inclination of the vehicle θ'_{ti}, which is determined from s'_{fi} and s'_{ri} increases gradually with increasing i_d as shown in Figure 6.14. The eccentricity e_i decreases with i_d for small magnitudes of slip ratio and takes a minimum value of -0.193 at $i_d = 51\%$. After this it increases again with i_d. The eccentricity e_i becomes less than $-1/6$ for the range of $i_d = 29 \sim 67\%$.

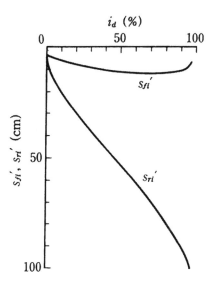

Figure 6.13. Relationship between amount of sinkage, of front-idler S'_{fi} rear sprocket S'_{ri} and slip ratio i_d (during driving action).

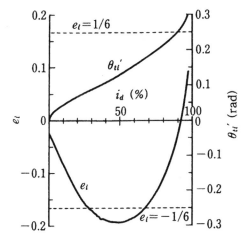

Figure 6.14. Relationship between angle of inclination of vehicle θ'_{ti} eccentricity of ground reaction e_i and slip ratio i_d (during driving action).

Figure 6.15 plots the relationships between the various energy components and the slip ratio. That is the effective input energy E_1, compaction energy E_2, slippage energy E_3, effective driving force energy E_4 are plotted as a function of the slip ratio i_d. The effective input energy E_1 moves to a constant value with increasing values of i_d, but the compaction energy E_2 follows a Hump type curve with a maximum value. The slippage energy E_3 increases almost linearly with i_d. The effective driving force energy E_4 takes a maximum value of 1471 kN cm/s at an optimum slip ratio of $i_{dopt} = 10\%$. After this it decreases gradually to a minimum value at $i_d = 71\%$, then increases again.

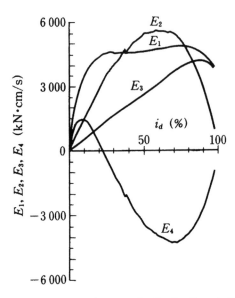

Figure 6.15. Relationship between energy elements E_1, E_2, E_3, E_4 and slip ratio i_d (during driving action).

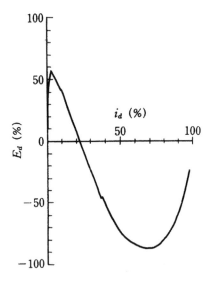

Figure 6.16. Relationship between tractive power efficiency E_d and slip ratio i_d (during driving action).

Figure 6.16 plots the relationship between tractive power efficiency E_d and slip ratio i_d. The tractive power efficiency E_d takes a maximum value of 57.0% at $i_d = 3\%$ after that it decreases rapidly with i_d to take on a minimum value of -86.9% at $i_d = 68\%$. After this minimum, it then increases again.

The conclusions here are, that when this bulldozer is running on this silty loam terrain during driving action and operating at $i_{dopt} = 10\%$ under with maximum driving force

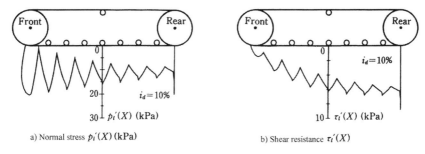

a) Normal stress $p_i'(X)$ (kPa)   b) Shear resistance $\tau_i'(X)$

Figure 6.17. Contact pressure distribution under flexible track belt (during driving action).

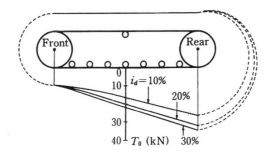

Figure 6.18. Distribution of track belt tension T_0 (during driving action).

energy, $T_{1opt} = 35.1$ kN and $T_{4opt} = 16.3$ kN can be developed at $s_{fi}' = 6.0$ cm, $s_{ri}' = 19.8$ cm, $\theta_{ti}' = 0.043$ rad, $e_i = -0.082$ and $E_d = 41.8\%$.

The contact pressure distribution acting on the flexible track belt of this bulldozer varies with the slip ratio i_d. As an example, Figure 6.17 shows the distribution of normal stress $p_i'(X)$ and shear resistance $\tau_i'(X)$ at $i_d = 10\%$. As shown in the diagram, there are several stress concentrations just under the road rollers and the contact pressure between the road rollers decreases due to the deflection of the flexible track belt. That is, the dynamic load acts repeatedly on the terrain. The pressure $p_i'(X)$ tends to increase toward the forward part of the track belt due to the negative eccentricity of the ground reaction. As will be discussed later, the amount of deflection of the front part of flexible track belt increases due to the relatively small track belt tension so that the amplitude of the contact pressure increases at the front part of the track belt. In contrast, $\tau_i'(X)$ shows the same wavy distribution, but, at the front part of the track belt, the shear resistance $\tau_i'(X)$ does not develop to any great degree because of the small amount of slippage.

Figure 6.18 shows the distributions of track belt tension T_0 around the track belt at $i_d = 10$, 20 and 30%. The track belt tension increases toward the rear part of the flexible track belt with i_d, while it equals the initial track belt tension H_0 at the part of the front-idler. At the base part of the rear sprocket, T_0 reaches 34.5 kN at $i_d = 30\%$. Furthermore, the amount of deflection of the track belt corresponds to the distribution of the track belt tension T_0. For instance, the track belt at the base of the rear sprocket is tensioned to an amount of deflection of 3.1 mm while the amount of deflection at the base of the front-idler is 7.1 mm.

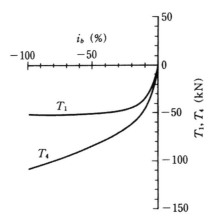

Figure 6.19. Relationship between braking force T_1 effective braking force T_4 and skid (during braking action).

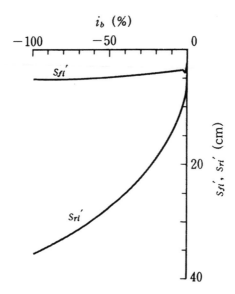

Figure 6.20. Relationship between amount of sinkage of front-idler s'_{fi} rear sprocket s'_{ri} and skid i_b (during braking action).

(b)   In the braking state
In the same way as has already been discussed, variations of braking force T_1, effective braking force T_4, amount of sinkage of the front-idler s'_{fi} and the rear sprocket s'_{ri}, angle of inclination of vehicle θ'_{ti}, eccentricity of ground reaction e_i, energy components $E_1 \sim E_4$, braking power efficiency E_b during braking action, can be calculated as functions of the skid i_b.

In terms of analysing the results, Figure 6.19 plots the relations between T_1, T_4 and the skid i_b. $|T_1|$ increases rapidly with the increasing $|i_b|$ for the lower ranges of skid, but it tends gradually to a constant value for the upper ranges of skid. On the other hand, $|T_4|$ increases gradually with increments in $|i_b|$. It is always larger than $|T_1|$ because of the increasing compaction resistance T_2 that occurs with increasing $|i_b|$.

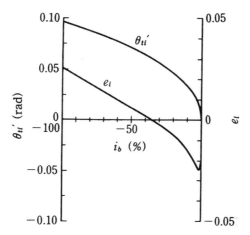

Figure 6.21. Relationship between angle of inclination of vehicle θ'_{ti}, eccentricity of ground reaction e_t and skid i_b (during braking action).

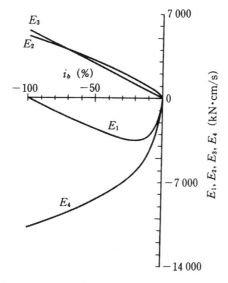

Figure 6.22. Relationship between energy elements E_1, E_2, E_3, E_4 and skid i_b (during braking action).

Figure 6.20 plots the relations between the sinkages s'_{fi}, s'_{ri} and the skid i_b. The rear sprocket sinkage s'_{ri} increases gradually with increments in $|i_b|$ in company with increased amounts of slip sinkage. In contrast s'_{fi} takes on an almost constant value. As a consequence, the angle of inclination of the vehicle θ'_{ti}, which is determined from s'_{fi} and s'_{ri}, increases parabolically with increases in $|i_b|$ as shown in Figure 6.21. The eccentricity e_i increases almost linearly with $|i_b|$ with a negative value to positive value transition at $i_b = -36\%$.

Figure 6.22 plots the relations between the various energy components and the skid. That is the effective input energy E_1, the compaction energy E_2, the slippage energy E_3, the effective braking force energy E_4 are plotted as a function of the skid i_b. $|E_1|$ decreases gradually to zero at $i_b = -100\%$ after taking on a maximum value of 3525 kN cm/s at $i_{bopt} = -19\%$.

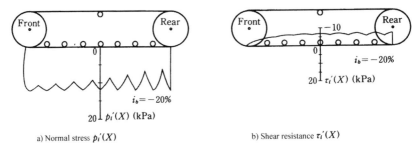

a) Normal stress $p_i'(X)$ b) Shear resistance $\tau_i'(X)$

Figure 6.23. Contact pressure distribution under flexible track belt (during braking action).

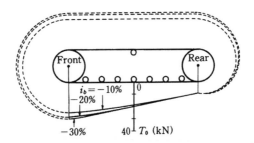

Figure 6.24. Distribution of track belt tension T_0 (during braking action).

The effective braking force energy $|E_4|$ increase parabolically while E_2 and E_3 increases gradually and almost linearly with increasing values of $|i_b|$.

In summary, when this bulldozer is running on this silty loam terrain and is operating during braking action at $i_{bopt} = -19\%$ under the maximum effective input energy, $T_{1opt} = -43.6$ kN and $T_{4opt} = -60.5$ kN can be developed at $s_{fi}' = 4.1$ cm, $s_{ri}' = 18.5$ cm, $\theta_{ti}' = 0.045$ rad, $e_i = -0.008$ and $E_b = 172\%$. The contact pressure distribution that acts on the flexible track belt varies with the skid i_b. To illustrate, Figure 6.23 shows the distributions of the normal stress $p_i'(X)$ and the shear resistance $\tau_i'(X)$ at $i_b = -20\%$. As shown in the diagram, there are several stress concentrations just under the road rollers and the contact pressure between the road rollers decreases due to the deflection of the flexible track belt. As will be discussed later, the amount of deflection of the rear part of the flexible track belt increases due to the relatively small track belt tension so that the amplitude of the contact pressure increases at the rear part of the track belt. On the other hand, $\tau_i'(X)$ shows a negative wavy distribution in the same way, but, at the front part of the track belt, the shear resistance $\tau_i'(X)$ does not develop to any great extent because of the small amount of slippage.

Figure 6.24 shows the distributions of track belt tension T_0 around the track belt at $i_b = -10$, -20 and -30%. The track belt tension increases toward the front part of the flexible track belt with $|i_b|$, whilst it equals the initial track belt tension H_0 at the part of the rear sprocket. At the base-part of the front-idler, T_0 reaches 29.9 kN at $i_b = -30\%$. Further, the track belt at the part of the front-idler is tensioned to an amount of deflection of 1.6 mm while the amount of deflection at the base part of the rear sprocket is 3.3 mm.

(2)   *Size effect of vehicle*
Since the size effect of a track model plate on the tractive performances of a bulldozer running on a clayey terrain could be enormous, the problem of the size effect of vehicle

needs be considered seriously in any investigations of the trafficability of an actual flexible tracked vehicle as mentioned in the previous Section 4.3.4.

In the section that follows, several influences of the vehicle dimensions on the tractive performance are analysed when several bulldozers having various sizes of track belt and vehicle weight are running under traction on a flat silty loam terrain. These studies can comprise a form of sensitivity analysis to machine scale.

Table 6.3 shows the vehicle dimensions of actual bulldozers and the terrain-track system constants calculated from the previous Eqs. (4.36) $\sim$ (4.45) for a size ratio of track model plate of $N = 4$, 8, 12 and 16 respectively.

Using the flow chart to calculate the tractive performance of the flexible tracked vehicle as shown in Figure 6.5 and using this input data, the relationships between the effective tractive effort T_4, the amount of sinkage of the rear sprocket s'_{ri}, the angle of inclination of the vehicle θ'_{ti}, the eccentricity of the ground reaction e_i and the slip ratio i_d have been mathematically simulated.

As to results, Figure 6.25 shows the computed effects of the size ratio N of the vehicle on the relations between the coefficient of traction i.e. the effective tractive effort T_4 divided by the vehicle weight W and the slip ratio i_d. For a small bulldozer of $N = 4$, the coefficient of traction takes a maximum value of $(T_4/W)_{max} = 0.824$ at $i_d = 38\%$. After that it becomes zero at $i_d = 94\%$. In contrast, for a large bulldozer of $N = 16$, the coefficient of traction takes on a maximum value of $(T_4/W)_{max} = 0.105$ at $i_d = 4\%$ but the bulldozer can not work at $i_d = 8\%$ onwards.

From this work, it is clearly demonstrated that the coefficient of traction of a bulldozer running on a soft terrain decreases remarkably with increases of size and weight of the bulldozer.

For this case, the tractive performance of a bulldozer having the same average contact pressure of 7.81 kPa will drop down from a size ratio of $N = 16$ onward.

Figure 6.26 shows the effects of the size ratio N of the vehicle on the relations between the relative amount of sinkage i.e. the amount of sinkage of the rear sprocket s'_{ri} divided by the track width B and the slip ratio i_d. As the relative amount of sinkage s'_{ri}/B tends to increase rapidly with increasing values of size ratio N for whole the range of slip ratio i_d, the land locomotion resistance is considered to increase remarkably with the size of the track belt of the vehicle.

Figure 6.27 shows the relations between the angle of inclination of vehicle θ'_{ti} and the slip ratio i_d. The angle θ'_{ti} increases with increases in the slip ratio i_d and with increases in the size ratio N. This means that the vehicle will tend to incline remarkably with the increases in the size of the track belt. The eccentricity of the ground reaction e_i tends to change from a negative value to a positive one at some slip ratio i_d, as shown in Figure 6.28. The absolute value of the minimum eccentricity e_{imin} tends to increase with increasing N. For a vehicle of size ratio $N = 16$, eccentricity e_i reduces down to less than $-1/6$. As a consequence the vehicle becomes unstable in the range of slip ratios of $i_d = 23 \sim 27\%$. Further, it is confirmed that the tractive power efficiency at the optimum slip ratio also decreases with increases in the size ratio N of the vehicle.

From the above simulation analytical results, it should be noticed that the coefficient of traction and the tractive power efficiency at the optimum slip ratio of a bulldozer operating on a weak silty loam terrain decrease remarkably, even if the average contact pressure is the same, with increases in the size of vehicle due to a diminution of the terrain-track system constants sensitivity to the size effect of the track belt.

Table 6.3. Dimensions of bulldozer running on soft terrain and terrain-track system constants.

Group	Description	Symbol (unit)				
Vehicle	Vehicle size ratio	N	4	8	12	16
	Vehicle weight	W (kN)	12.5	50.0	112.5	200.0
	Track width	B (cm)	50	100	150	200
	Contact length	D (cm)	160	320	480	640
	Average contact pressure	p_m (kPa)	7.81	7.81	7.81	7.81
	Radius of front-idler	R_l (cm)	15	30	45	60
	Radius of rear sprocket	R_r (cm)	15	30	45	60
	Interval of road roller	R_p (cm)	20	40	60	80
	Radius of road roller	R_m (cm)	4	8	12	16
	Height of centre of gravity of vehicle	h_o (cm)	30	60	90	120
Dimensions	Distance of application point of effective driving force from central axis of vehicle	I_d (cm)	150	300	450	600
	Height of application point of effective driving force	h_d (cm)	25	50	75	100
	Grouser height	H (cm)	6	12	18	24
	Grouser pitch	G_p (cm)	18	36	54	72
	Initial track belt tension	H_o (kN)	2.45	9.80	22.05	39.20
	Eccentricity of center of gravity of vehicle	e		0.00		
	Circumferential speed of track belt	V' (cm/s)		100		
Terrain-track system constants — Loading test		k_1 (N/cm^{n_1+2})	2.87	1.26	7.76×10^{-1}	5.51×10^{-1}
		N_1	9.41×10^{-1}	1.07	1.15	1.21
		k_2 (N/cm^{n_2+2})	5.57	5.52	5.48	5.46
		n_2	8.36×10^{-1}	8.95×10^{-1}	9.31×10^{-1}	9.58×10^{-1}
		m_c (kPa)	4.09	3.36	3.00	2.76
Traction test		m_f	3.42×10^{-1}	3.11×10^{-1}	2.94×10^{-1}	2.82×10^{-1}
		(1/cm)	1.39×10^{-1}	7.84×10^{-2}	5.61×10^{-2}	4.43×10^{-2}
		c_0 $(cm^{2c_1-c_2+1}/N^{c_1})$	2.72×10^{-1}	6.93×10^{-1}	1.20	1.76
		c_1	7.34×10^{-1}	5.86×10^{-1}	5.14×10^{-1}	5.46×10^{-1}
		c_2	3.80×10^{-1}	4.78×10^{-1}	5.46×10^{-1}	6.01×10^{-1}

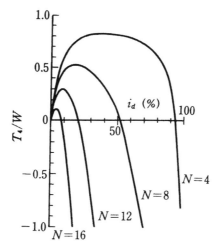

Figure 6.25. Relationship between coefficient of traction T_4/W and slip ratio i_d for various size ratios N of vehicle.

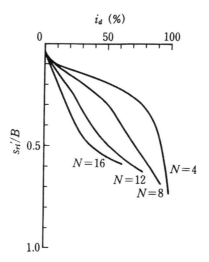

Figure 6.26. Relationship between relative amount of sinkage s'_{ri}/B and slip ratio i_d for various size ratio N of vehicle.

(3)   Effect of initial track belt tension

It is a well-established empirical fact that the tractive performance of a bulldozer running on a soft terrain depends to a very large degree on the flexibility of the track belt. Usually, the track belts on working machines are tensioned by use of an adjuster composed of a cylinder and a spring that is mounted on the axle of the front-idler. Typically, the initial track belt tension is set up at more than 20% of the vehicle weight by use of a grease gun. However, under normal operations, the track belt tension reduces due to wear in the undercarriage parts of the bulldozer i.e. in the front-idler, track link, pin, bush, road roller, rear sprocket and so on. In general, the tractive effort of a bulldozer decreases with reductions in the

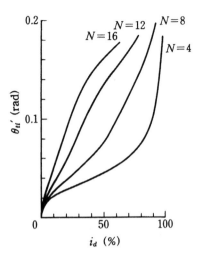

Figure 6.27. Relationship between angle of inclination of vehicle θ'_{ti} and slip ratio i_d for various size ratio N of vehicle.

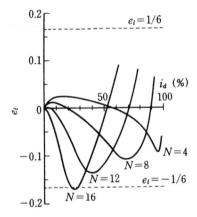

Figure 6.28. Relationship between eccentricity of ground reaction e_i and slip ratio i_d for various size ratio N of vehicle.

initial track belt tension because of an increasing land locomotion resistance with increases in amounts of sinkage [26, 27].

In this section, several effects of the initial track belt tension on the optimum effective tractive effort, the amount of sinkage of the rear sprocket, the eccentricity of the ground reaction, the angle of inclination of the vehicle and the tractive power efficiency at the optimum slip ratio of a bulldozer of vehicle weight 150 kN are forecast through the use of a rigorous mathematical simulation analysis program.

We assume the bulldozer to be simulated is equipped with a flexible track belt and to be operating on a flat weak remolded silty loam terrain. The track belt is equipped with trapezoidal rubber grousers of height 6 cm, pitch 18 cm, contact length 2.0 cm and base length 8.9 cm. The terrain-track system constants that prevail between the track belt and a silty loam terrain of water content 30% and cone index 30 kPa are summarized in Table 6.4.

Table 6.4. Terrain-track system constants for track belt equipped with trapezoidal rubber grousers.

Track plate loading and unloading test

$p \le 16.8$ (kPa)	$k_1 = 2.255$ $n_1 = 1.120$
$p > 16.8$ (kPa)	$k_2 = 6.669$ $n_2 = 5.938 \times 10^{-1}$
	$k_3 = k_4 = 30.03$ $n_3 = n_4 = 0.632$

Track plate traction test

$m_c = 3.626$ (kPa) $m_f = 0.356$ $a = 0.148$ (1/cm)

Track plate slip test

$c_0 = 0.253$ $c_1 = 0.751$ $c_2 = 0.360$

Table 6.5. Dimensions of tracked bulldozer equipped with trapezoidal rubber grousers running on soft terrain.

Vehicle weight	W (kN)	150
Track width	B (cm)	150
Contact length	D (cm)	320
Average contact pressure	p_m (kPa)	15.6
Radius of front-idler	R_f (cm)	35
Radius of rear sprocket	R_t (cm)	35
Grouser height	H (cm)	6
Grouser pitch	G_p (cm)	18
Interval of road roller	R_p (cm)	40
Eccentricity of center of gravity of vehicle	e	−0.05
		0
		0.05
Distance of application point M of effective driving force from central axis of vehicle	l_d (cm)	310
Height of application point of effective driving force	h_d (cm)	40
Initial track belt tension	H_0 (kN)	$0 \sim 50$
Circumferential speed of track belt (during driving state)	V' (cm/s)	100
Height of center of gravity of vehicle	h_g (cm)	70

The dimensions and specifications of the bulldozer are shown in Table 6.5. For this system a mathematical simulation has been run for three values of eccentricity of the center of gravity of the bulldozer e, namely $e = -0.05$, 0 and 0.05 and for ten kinds of initial track belt tension H_0, namely $H_0 = 0$, 2, 4, 6, 8, 10, 20, 30, 40 and 50 kN. Substituting these values as input data into the flow chart of Figure 6.5, the effects of the initial track belt tension on the tractive performances of the bulldozer may be analysed [28].

Figure 6.29 graphs the relationship between the optimum effective tractive effort T_{4opt} and the initial track belt tension H_0 for various values of eccentricity of the center of gravity of vehicle e. It is especially evident that T_{4opt} drops down with decreasing values of H_0 from $H_0 = 10$ kN for $e = 0.00$ or 0.05.

Figure 6.30 shows the relationship between the amounts of sinkage of the rear sprocket s'_{ri} at the optimum slip ratio and H_0 for $e = -0.05$, 0.00 and 0.05. The rear sprocket sinkage s'_{ri} increases rapidly with decreasing values of H_0 from $H_0 = 10$ kN for each of the values of e.

Figure 6.31 shows the relationship between the angle of inclination of the vehicle θ'_{ti} at the optimum slip ratio and H_0 for e values of −0.05, 0.00 and 0.05 respectively. The

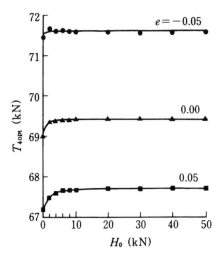

Figure 6.29. Relationship between optimum effective driving force T_{4opt} and initial belt tension H_0 for various kinds of eccentricity e of center of gravity of vehicle.

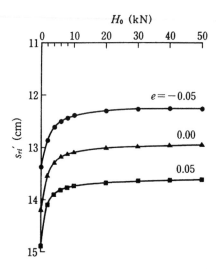

Figure 6.30. Relationship between amount of sinkage of rear sprocket s'_{ri} and initial track belt tension for various kinds of eccentricity e of center of gravity of vehicle.

inclination θ'_{ti} decreases gradually with decreasing values of H_0 and then decreases rapidly for H_0 values less than 10 kN for each setting of e, where the vehicle can move with stability.

Figure 6.32 shows the relationship between the eccentricity e_i of the ground reaction at the optimum slip ratio and H_0 for various ranges of e. The eccentricity e_i is almost a constant value independent of changes in H_0. This applies for each value of e.

Figure 6.33 shows the relationship between the tractive power efficiency E_d at the optimum slip ratio and H_0 for $e = -0.05$, 0.00 and 0.05. From this diagram, it is clear that E_d decreases rapidly with decreasing values of H_0 less than 10 kN for each value of e.

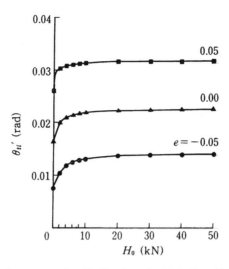

Figure 6.31. Relationship between angle of inclination of vehicle θ'_{ti} and initial track belt tension H_0 for various kinds of eccentricity e of center of gravity of vehicle.

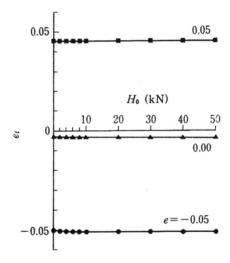

Figure 6.32. Relationship between eccentricity e_i of ground reaction and initial track belt tension H_0 for various e ($i_d = i_{dopt}$).

From the results of the above simulation analysis, it is evident that, for the various eccentric- ities of the center of gravity of the vehicle $e = -0.05$, 0.00 and 0.05, the optimum effective tractive effort, the angle of inclination of vehicle and the tractive power efficiency at the optimum slip ratio decreases rapidly with decreasing values of the initial track belt tension. However, the amount of sinkage of the rear sprocket at the optimum slip ratio increases rapidly with decreasing values of the initial track belt tension.

In contrast to this situation, a rigid tracked vehicle equipped with a rubber track belt having a great initial track belt tension can develop a larger optimum effective tractive

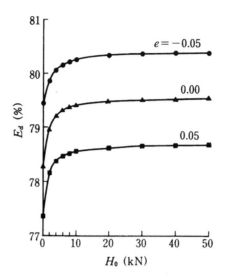

Figure 6.33. Relationship between tractive power efficiency E_d and initial track belt tension H_0 for various e ($i_d = i_{dopt}$).

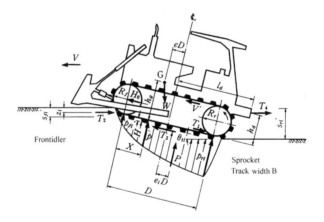

Figure 6.34. Forces acting on a bulldozer running on decomposed granite sandy terrain.

effort and tractive power efficiency over and above those of a flexible tracked vehicle running on a soft terrain as discussed above.

In this case, it is necessary to control the initial track belt tension to be always larger than 10 kN to maintain a sensible operation of a bulldozer running on a soft terrain.

6.6.2 Decomposed granite sandy terrain

In this section we will consider a simulation of the traffic performance of a flexible tracked vehicle, such as a bulldozer, running on a flat terrain composed of an accumulated decomposed weathered granite sandy soil as already has been discussed in Section 4.3.2.

Figure 6.34 shows the system of forces acting on the bulldozer. The structure of the track belt is that of a flexible belt equipped with equilateral trapezoidal rubber grousers of base

Table 6.6. Dimensions of small bulldozer.

Vehicle weight	W (kN)	50
Track width	B (cm)	25
Contact length	D (cm)	320
Average contact pressure	p_m (kPa)	31.25
Radius of frontidler	R_f (cm)	50
Radius of rear sprocket	R_t (cm)	50
Grouser height	H (cm)	6.5
Grouser pitch	G_p (cm)	14.6
Interval of road roller	R_p (cm)	40
Radius of road roller	R_m (cm)	8
Eccentricity of center of gravity of vehicle	e	−0.02
Distance of application point M effective driving force from central axis of vehicle	l_d (cm)	300
Height of application point of effective driving force	h_d (cm)	60
Initial track belt tension	H_o (kN)	19.6
Circumferential speed of track belt (during driving state)	V' (cm/s)	100
Height of center of gravity of vehicle	h_g (cm)	100

length $L = 2$ cm, grouser height $H = 6.5$ cm, grouser pitch $G_p = 14.6$ cm and trim angle $\alpha = \pi/6$ rad as shown in the previous Figure 4.6. The terrain-track system constants that operate between the decomposed granite sandy soil of dry density 1.44 g/cm^3 and a given track-model-plate of width $B = 25$ cm are $k_1 = 8.526$ N/cm$^{n_1+2}$, $n_1 = 0.866$, $m_c = 0$ kPa, $m_f = 0.769$, $a = 0.244$ cm^{-1}, $c_0 = 1.588$ cm$^{2c_1-c_2+1}$/N^{c_1}, $c_1 = 0.075$ and $c_2 = 0.240$ as given in the previous Table 4.2.

Table 6.6 shows the specifications and dimensions of a small bulldozer of weight 50 kN, width of track belt 25 cm and contact length of track belt 320 cm. In this case, it is not necessary to consider the size effect of the terrain-track system constants because the width of the track belt is the same as that of the track-model-plate.

(1) At driving state

By using the above terrain-track system constants and vehicle dimensions as the input data to the flow chart of Figure 6.5, the variations of driving force T_1, effective driving force T_4, amounts of sinkage of the front-idler s'_{fi} and rear sprocket s'_{ri}, angle of inclination of the vehicle θ'_{ti}, the eccentricity of the ground reaction e_i, the various energy components $E_1 \sim E_4$ and the tractive power efficiency E_d can be calculated as a function of the slip ratio i_d for a vehicle under driving action.

Figure 6.35 portrays the relations between T_1, T_4 and i_d resulting from the simulation calculations. The driving force T_1 increases rapidly to a constant value with increasing values of i_d. The effective driving force T_4 decreases gradually with i_d after peaking at a maximum value of $T_{4max} = 39.7$ kN at $i_d = 23\%$.

Figure 6.36 shows the relations between s'_{fi}, s'_{ri} and i_d. The Figure shows both the sinkages s'_{fi}, s'_{ri} increasing gradually with i_d. The sinkage s'_{ri} is always larger than s'_{fi} because of the increasing amount of slip sinkage at the rear sprocket. Thus, it can be seen that the angle of inclination of the vehicle θ'_{ti} which may be calculated from s'_{fi} and s'_{ri} increases parabolically with increasing values of slip ratio i_d. The phenomenon is illustrated in Figure 6.37. As shown in the diagram, the eccentricity of the ground reaction e_i has a negative value. It decreases gradually after it passes a maximum value of −0.0108 at $i_d = 5\%$.

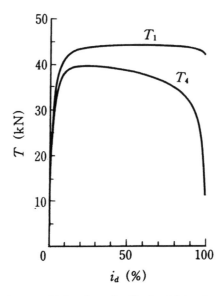

Figure 6.35. Relationship between driving force T_1 effective driving force T_4 and slip ratio i_d.

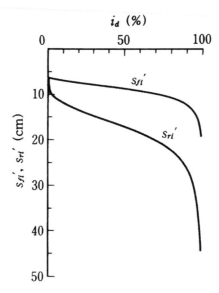

Figure 6.36. Relationship between amount of sinkage of frontidler s'_{fi} rear sprocket s'_{ri} and slip ratio i_d.

Figure 6.38 plots the relationships that exist between the various energy components and the slip ratio i_d. Thus, Figure 6.38 shows the effective input energy E_1, the compaction energy E_2, the slippage energy E_3 and the effective tractive effort energy E_4 as a function of the slip ratio i_d. The effective input energy E_1 increases quite rapidly and then tends to a constant value with i_d.

The energies E_2 and E_4 follow a hump-type curve having a maximum value. The component E_3 increases almost linearly with increasing values of i_d. E_4 peaks at a maximum

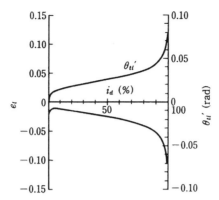

Figure 6.37. Relationship between angle of inclination of vehicle θ'_{ti} eccentricity of ground reaction and slip ratio i_d (during driving action).

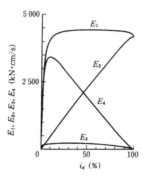

Figure 6.38. Relationship between energy elements E_1, E_2, E_3, E_4 and skid i_d (during driving action).

value of 3430 kNcm/s at $i_d = 10\%$ and after that it decreases almost linearly to zero at $i_d = 100\%$.

Figure 6.39 shows the relations between the tractive power efficiency E_d and the slip ratio i_d. E_d has a sharp peak with a maximum value of 90.4% at $i_d = 2\%$. After that, it decreases, essentially linearly, to zero at $i_d = 100\%$.

In summary, when this small bulldozer is running on a decomposed weathered granite sandy terrain during driving action and is operating at $i_{dopt} = 10\%$ under maximum effective tractive effort energy, $T_{1opt} = 41.1$ kN and $T_{4opt} = 38.1$ kN can be developed at $s'_{fi} = 6.9$ cm, $s'_{ri} = 11.7$ cm, $\theta'_{ti} = 0.015$ rad, $e_i = -0.0118$ and $E_d = 83.5\%$.

The contact pressure distribution acting on the flexible track belt of a small bulldozer varies with the slip ratio i_d. For example, Figure 6.40 shows the distribution of the normal stress $p'_i(X)$ and the shear resistance $\tau'_i(X)$ for $i_d = 10 \sim 90\%$ for the simulated system. The normal stress $p'_i(X)$ tends to increase toward the front part of the track belt due to the negative value of e_i. The amount of deflection of the track belt and the amplitude of the wavy distribution of the normal stress on the front part of the flexible track belt become large due to the relatively small track belt tension. On the other hand, the shear resistance $\tau'_i(X)$ also shows a wavy distribution, but the shear resistance does not develop significantly at the front part of the track belt for $i_d = 10\%$ because the amount of slippage is very small.

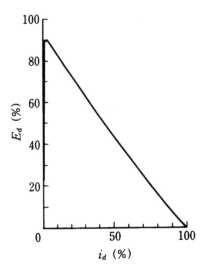

Figure 6.39. Relationship between tractive power efficiency E_d and slip ratio i_d (during driving action).

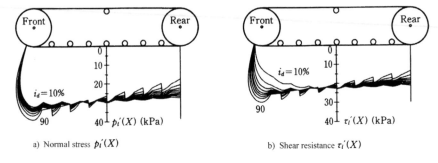

a) Normal stress $p_i'(X)$

b) Shear resistance $\tau_i'(X)$

Figure 6.40. Contact pressure distribution (during driving action).

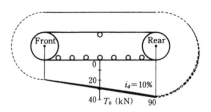

Figure 6.41. Distribution of track belt tension T_0 (driving state).

Figure 6.41 shows the distribution of the track belt tension T_0 around the track belt for $i_d = 10 \sim 90\%$. The track belt tension increases toward the rear part of the flexible track belt with i_d, whilst it shows an initial track belt tension of $H_0 = 19.6\,$kN at the base of the front-idler. At the base of the rear sprocket, T_0 reaches $36.2\,$kN at $i_d = 10\%$. The track belt at the base of the rear sprocket is tensioned to provide a deflection of $0.43\,$mm while the magnitude of the deflection at the base of the front-idler is $1.96\,$mm for $i_d = 10\%$.

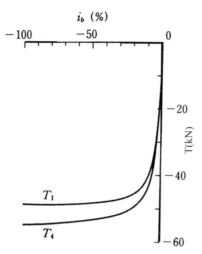

Figure 6.42. Relationship between braking force T_1 effective braking force T_4 and skid i_b.

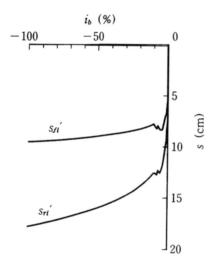

Figure 6.43. Relationship between amount of sinkage of frontidler s'_{fi} rear sprocket s'_{ri} and skid i_b.

(2) *At braking state*

In the same manner as has been already described, the variations of the braking force T_1, the effective braking force T_4, the amounts of sinkage of the front-idler s'_{fi} and rear sprocket s'_{ri}, the angle of inclination of the vehicle θ'_{ti}, the eccentricity of the ground reaction e_i, the various energy components $E_1 \sim E_4$ and the braking power efficiency E_b as a function of the skid i_b during braking action, can be calculated by use of the flow chart of Figure 6.5. In terms of quantitative results, Figure 6.42 shows the relations between T_1, T_4 and i_b. Both the values of $|T_1|$, $|T_4|$ increase with increasing values of $|i_b|$ and tend to constant values. $|T_4|$ is always greater than $|T_1|$ for all values of skid.

Figure 6.43 shows the relations between the sinkages s'_{fi}, s'_{ri} and i_b. Both the values of s'_{fi} and s'_{ri} increase gradually with increasing values of $|i_b|$. The rear sinkage s'_{ri} is always

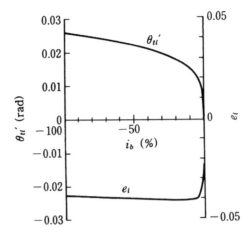

Figure 6.44. Relationship between angle of inclination of vehicle θ'_{ti} eccentricity of ground reaction e_i and skid i_b (during braking action).

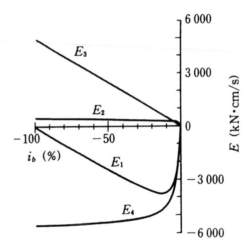

Figure 6.45. Relationship between energy elements E_1, E_2, E_3, E_4 and skid i_b (during braking action).

larger than the front sinkage s'_{fi} due to the amount of slip sinkage that occurs at the rear sprocket.

Figure 6.44 shows the relations between θ'_{ti}, e_i and i_b. The inclination θ'_{ti} increases parabolically with increasing values of $|i_b|$. On the other hand, the eccentricity e_i takes on a negative value which increases slightly after it takes a minimum value of -0.0409 at a skid $i_b = -8\%$.

Figure 6.45 shows the relations between the various energy elements – namely the effective input energy E_1, the compaction energy E_2, the slippage energy E_3, the effective braking force energy E_4 and the skid i_b. The element $|E_1|$ takes a maximum value of 3782 kNcm/s at an optimum skid i_{bopt} of -13%. It then decreases gradually to zero at $i_b = -100\%$. The compaction energy E_2 and the slippage energy E_3 increases almost linearly with increasing values of $|i_b|$, but $|E_4|$ increases parabolically and tends to a constant value with $|i_b|$.

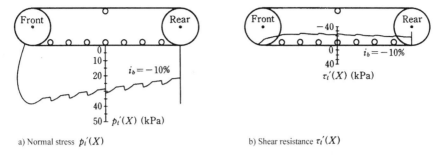

a) Normal stress $p_i'(X)$ b) Shear resistance $\tau_i'(X)$

Figure 6.46. Contact pressure distribution (during braking action).

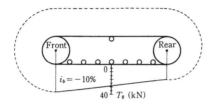

Figure 6.47. Distribution of track belt tension T_0 (braking state).

In summary, when a simulated small bulldozer is in running operations on a decomposed weathered granite sandy soil during braking action and is operating at $i_b = -13\%$ under a maximum effective input energy, $T_{1opt} = -43.5$ kN and $T_{4opt} = -47.0$ kN can be developed at $s_{fi}' = 7.9$ cm, $s_{ri}' = 13.0$ cm, $\theta_{ti}' = 0.016$ rad, $e_i = -0.041$ and $E_b = 124\%$.

The contact pressure distribution acting on the flexible track belt of a small bulldozer varies with the skid i_b. For example, Figure 6.46 shows the simulated distributions of the normal stress $p_i'(X)$ and the shear resistance $\tau_i'(X)$ at $i_b = -10\%$.

The normal stress $p_i'(X)$ tends to increase toward the front part of the track belt due to the negative value of e_i and there are several stress concentrations just under the track rollers. At the rear part of the track belt, both the amount of deflection of the track belt and the amplitude of the wavy distribution of the normal stress $p_i'(X)$ becomes large due to the relatively small track belt tension. On the other hand, the shear resistance $\tau_i'(X)$ shows a negative wavy distribution, but the shear resistance does not develop significantly at the front part of the track belt because the amount of slippage is very small.

Figure 6.47 shows the distribution of the track belt tension T_0 around the track belt at $i_b = -10\%$. The track belt tension increases toward the front part of the flexible track belt, whilst it shows an initial track belt tension of $H_0 = 19.6$ kN at the base part of the rear sprocket. At the base of the front-idler, T_0 reaches 36.8 kN at $i_b = -10\%$. The track belt at the base part of the front-idler is tensioned such as to provide an amount of deflection of 0.1 mm while the amount of deflection at the part of the rear sprocket is 3.8 mm for $i_b = -10\%$.

6.6.3   Snow covered terrain

The tractive performance during driving action and the braking performance during braking action of a flexible tracked over-snow vehicle of weight 40.0 kN running on a snow covered terrain can be forecast through use of a mathematical model based simulation process that follows the procedures shown in the flow chart of Figure 6.5.

Table 6.7. Specifications of test vehicle.

Vehicle weight	W (kN)	40
Track width	B (cm)	180
Contact length	D (cm)	480
Average contact pressure	p_m (kPa)	2.31
Radius of front-idler	R_f (cm)	25
Radius of rear sprocket	R_r (cm)	25
Grouser height	H (cm)	3
Grouser pitch	G_p	9
Eccentricity of center of gravity	e	0.00
Height of center of gravity	h_g (cm)	95
Distance of application point from axis	l_b (cm)	300
Distance of application point from axis	L_d (cm)	300
Height of application point of force	h_b (cm)	50
Height of application point of force	h_d (cm)	50
Initial track belt tension	H_0 (kN)	29.4
Circumferential speed of track (driving)	V' (cm/s)	100
Vehicle speed (during braking state)	V (cm/s)	100

The specifications of the vehicle system to be simulated are shown in Table 6.7. The terrain-track system constants that develop between a track model plate equipped with standard T shaped grousers and a shallow deposited snow covered terrain of depth 20 cm are $k_2 = 0.315$ N/cm$^{n_2+1}$, $n_2 = 1.220$, $k_4 = 32.34$ N/cm$^{n_4+1}$, $n_4 = 0.862$ from track model plate loading and unloading test results, $f_s = 1.86$, $f_m = 0.01$ and $j_m = 1.5$ cm for hump-type shear deformation relations, and $c_0 = 0.685$ cm$^{2c_1-c_2+1}$/N^{c_1}, $c_1 = 0.694$ and $c_2 = 0.476$ for slip sinkage relations from track model plate traction test results.

(1)   At driving state

The variations of the driving force T_1, the effective driving force T_4, the amounts of sinkage of the front-idler s'_{fi} and the rear sprocket s'_{ri}, the angle of inclination of the vehicle θ'_{ti}, the eccentricity of ground reaction e_i, the various energy factors $E_1 \sim E_4$ and the tractive power efficiency E_d are a function of the slip ratio i_d can now be simulated.

   The data are for an over snow vehicle working under driving action and running on a flat snow covered terrain of $\beta = 0$ rad.

   Figure 6.48 shows the results of the simulation in terms of the relations between T_1, T_4 and i_d. Both the values of T_1 and T_4 take maximum values of $T_{1max} = 69.6$ kN and $T_{4max} = 68.6$ kN at $i_d = 0.3\%$ respectively. Also, both these factors decrease rapidly after taking a peak value. Figure 6.49 shows the relations between the sinkages s'_{fi}, s'_{ri} and i_d. The front sinkage s'_{fi} takes a minimum value of 3.1 cm at $i_d = 0.3\%$ and after that it increases gradually. The rear sinkage s'_{ri} increases parabolically with increasing values of slip ratio i_d. The rear sinkage s'_{ri} is always larger than the front sinkage s'_{fi} due to the increasing amount of slip sinkage that occur with i_d. As shown in Figure 6.50, the vehicle inclination θ'_{ti} reaches a peak value of 0.011 rad at $i_d = 0.5\%$ and the eccentricity e_i takes a maximum value of 0.0802 at $i_d = 0.3\%$. After these points both these factors decrease rapidly.

   Figure 6.51 shows the relations between the various energy components – namely the effective input energy E_1, the compaction energy E_2, the slippage energy E_3 and the effective tractive effort energy E_4 and the slip ratio i_d. The effective input energy E_1 and the effective tractive effort energy E_4 both have peak values at $i_d = 0.3\%$ and then

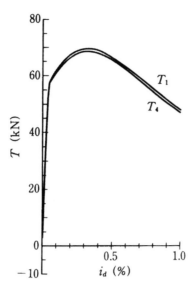

Figure 6.48. Relationship between driving force T_1 effective driving force T_4 and slip ratio i_d (driving state).

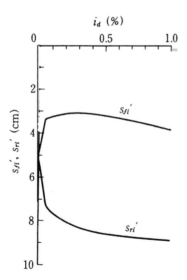

Figure 6.49. Relationship between amount of sinkage of frontidler s'_{fi} rear sprocket s'_{ri} and slip ratio i_d (driving state).

decrease rapidly. E_2 and E_3 increase linearly with i_d. E_4 has a peak value of 6843 kN cm/s at $i_{dopt} = 0.3\%$. Figure 6.52 shows the relationship between the tractive power efficiency E_d and slip ratio i_d. E_d takes a maximum value of 98.6% at $i_d = 0.05\%$ and after that it decreases gradually.

To summarise, when the simulated flexible tracked over-snow vehicle is running on a snow-covered terrain during driving action and is operating at $i_{dopt} = 0.3\%$ under the

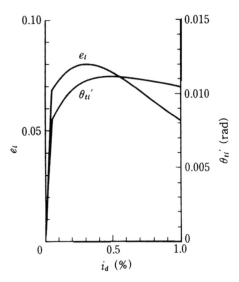

Figure 6.50. Relationship between angle of inclination of vehicle θ'_{ti} eccentricity of ground reaction e_i and slip ratio i_d (driving state).

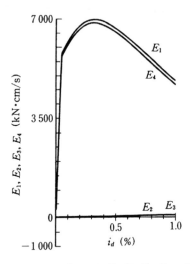

Figure 6.51. Relationship between energy elements E_1, E_2, E_3, E_4 and slip ratio i_d (driving action).

maximum effective tractive effort energy, $T_{1opt} = 69.6\,\text{kN}$ and $T_{4opt} = 68.6\,\text{kN}$ can be developed at $s'_{fi} = 3.1\,\text{cm}$, $s'_{ri} = 8.3\,\text{cm}$, $\theta'_{ti} = 0.011\,\text{rad}$, $e_i = 0.0802$ and $E_d = 98.4\%$.

The contact pressure distribution acting on the simulated flexible tracked over-snow vehicle varies with the slip ratio i_d. Figure 6.53 shows the distributions of normal stress $p'_i(X)$ and shear resistance $\tau'_i(X)$ at $i_d = 0.3$, 0.6 and 0.9%. As shown in the diagrams, there are no stress concentrations under the road rollers. This also shows that repetitive loading does not occur on the snow-covered terrain due to the deflection of the track belt between road rollers.

The normal stress $p'_i(X)$ tends to increase toward the rear part of the track belt due to the positive value of e_i. In spite of the large amount of deflection of the front part of the

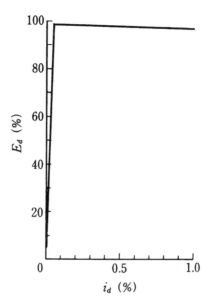

Figure 6.52. Relationship between tractive power efficiency E_d and slip ratio i_d (driving state).

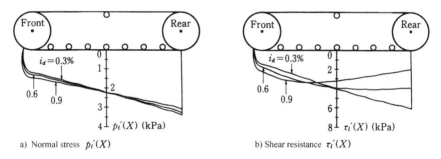

a) Normal stress $p_l'(X)$ b) Shear resistance $\tau_l'(X)$

Figure 6.53. Contact pressure distribution (during driving action).

track belt due to the relatively small track belt tension, $p_i'(X)$ does not show any major wavy distribution. This means that there is no unloading process on the contact part of the track belt that accompanies increasing values of the angle of inclination of the vehicle. On the other hand, $\tau_i'(X)$ shows a gentle wavy distribution and a large thrust is developed at the front part of the track belt.

Figure 6.54 shows the distributions of the track belt tension T_0 around the track belt at $i_d = 0.3$, 0.6 and 0.9%. The belt tension T_0 increases almost linearly toward the rear part of the track belt while the initial track belt tension of $H_0 = 29.4$ kN acts on the base part of the front-idler.

At the base part of the rear sprocket, the track belt tension reaches 62.9 kN at $i_d = 0.3\%$. The track belt at the base part of the rear sprocket is tensioned so as to provide a deflection of 1.6 mm while the amount of deflection at the base-part of the front-idler is 3.2 mm for $i_d = 0.3\%$.

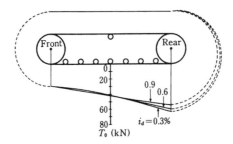

Figure 6.54. Distribution of track belt tension T_0 (during driving action).

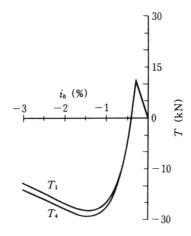

Figure 6.55. Relationship between braking force T_1 effective braking force T_4 and skid i_b.

(2)   *At braking state*
In the same manner as has already been worked through, the variations of the braking force T_1, the effective braking force T_4, the amounts of sinkage of the front-idler s'_{fi} and the rear sprocket s'_{ri}, the angle of inclination of the vehicle θ'_{ti}, the eccentricity of the ground reaction e_i, the various energy components $E_1 \sim E_4$ and the braking power efficiency E_b with variation in skid i_b during braking action of the over snow vehicle running on a flat snow covered terrain of $\beta = 0$ rad can be simulated.

Figure 6.55 shows the results of the simulation and plots the relations between T_1, T_4 for variation in i_b. The forces T_1 and T_4 have peak values of $T_{1max} = 11.6$ kN and $T_{4max} = 11.1$ kN at $i_b = -0.3\%$ respectively. After this peak they decrease rapidly to a minimum value of $T_{1min} = -27.4$ kN and $T_{4min} = -29.2$ kN at $i_b = -1.4\%$ respectively and after that they increase gradually.

Figure 6.56 graphs the relations between s'_{fi}, s'_{ri} and i_b. The front sinkage s'_{fi} decreases gradually with i_b after taking a maximum value of 8.4 cm at $i_b = -1.5\%$. The rear sinkage s'_{ri} increases parabolically with increasing values of $|i_b|$. The rear sinkage s'_{ri} is always larger than the front sinkage s'_{fi} due to the increasing amount of slip sinkage that occurs at the rear sprocket.

Also, as shown in Figure 6.57, the inclination θ'_{ti} increases with increasing values of $|i_b|$ and the eccentricity e_i takes a maximum value of 0.0138 at $i_b = -0.3\%$. After a peak

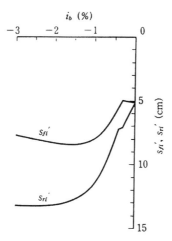

Figure 6.56. Relationship between amount of sinkage of frontidler s'_{fi} rear sprocket s'_{ri} and skid i_b (braking state).

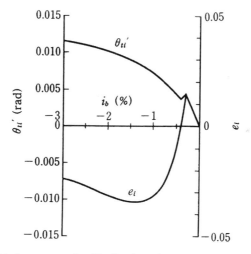

Figure 6.57. Relationship between angle of inclination of vehicle θ'_{ti} eccentricity of ground reaction e_i and skid i_b (braking state).

e_i moves rapidly to a minimum value of -0.0345 at $i_b = -1.4\%$. After this, it increases gradually.

Figure 6.58 shows the relations between the effective input energy E_1, the compaction energy E_2, the slippage energy E_3, the effective braking force energy E_4 and the skid i_b. Both the value of E_1 and E_4 take peak values at the same skid of $i_b = -0.3\%$. After that, they both drop down to minimum values at $i_b = -1.4\%$. Then, they increase gradually. The effective input energy $|E_1|$ takes a maximum value of 2706 kNcm/s at $i_{bopt} = -1.4\%$. The braking power efficiency E_b takes a minimum value of 34.4% after that it increases almost linearly with increasing values of $|i_b|$.

To summarise, when the particular simulated flexible tracked over-snow vehicle is running on a snow covered terrain during braking action and is operating at $i_{bopt} = -1.4\%$ under

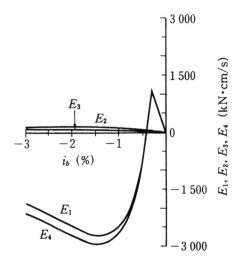

Figure 6.58. Relationship between energy elements E_1, E_2, E_3, E_4 and skid i_b (braking state).

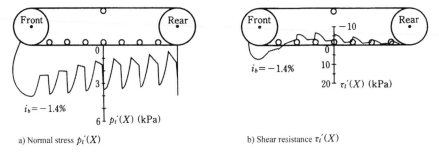

a) Normal stress $p_i'(X)$ b) Shear resistance $\tau_i'(X)$

Figure 6.59. Contact pressure distribution (during braking action).

maximum effective input energy, $T_{1opt} = -27.4$ kN and $T_{4opt} = -29.2$ kN can be developed at $s_{fi}' = 8.4$ cm, $s_{ri}' = 12.6$ cm, $\theta_{ti}' = 0.009$ rad, $e_i = -0.0345$ and $E_b = 108.0\%$. This shows that a repetitive loading occurs on the snow-covered terrain.

The contact pressure distribution acting on a flexible tracked over snow vehicle varies with skid i_b. For example, Figure 6.59 shows the simulated distributions of normal stress $p_i'(X)$ and shear resistance $\tau_i'(X)$ at $i_b = -1.4\%$. As shown in the diagram, there are stress concentrations that occur just under the road rollers and $p_i'(X)$ decreases between the road rollers due to the deflection of the track belt. The normal stress $p_i'(X)$ tends to increase toward to the front part of the track belt due to a negative value of e_i. At the rear part of the track belt, both the magnitude of the deflection of the track belt and the amplitude of the distribution of normal stress increase due to the relatively small value of the track belt tension. On the other hand, $\tau_i'(X)$ shows the same wavy distribution, but the value of shear resistance changes from positive value to negative one at some point on the track belt. After that, $|\tau_i'(X)|$ increases toward to the rear part of the track belt and then decreases gradually. This phenomenon is a special feature of snow covered terrain and it is evident that the drag is developed mainly on the front side of the middle part of the track belt.

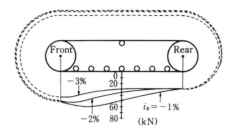

Figure 6.60. Distribution of track belt tension T_0 (braking state).

Figure 6.60 shows the distributions of the track belt tension T_0 around the track belt at $i_b = -1$, -2, and -3%. T_0 increases toward to the front part of the track belt whilst the initial track belt tension of $H_0 = 29.4$ kN acts on the base part of the rear sprocket. At the base part of the front-idler, the track belt tension reaches 50.0 kN at $i_b = -2\%$. The track belt at the base part of the front-idler is tensioned as to provide an amount of deflection of 0.3 mm whilst the amount of deflection at the base part of the rear sprocket is 2.55 mm at $i_b = -2\%$.

6.7   SUMMARY

In this chapter we have addressed one of the very difficult problems in everyday terramechanics, namely the problem of predicting the performance of a specific tracked-vehicle with flexible track system which is required to operate upon a particular complex terrain.

In an attempt to solve this problem in a generic manner, a modeling system based on a mixture of experimentally defined parameters and analytical considerations has been developed. By use of this modeling system and by means of a process of recursive analysis (using computers) it has been shown that the interplay between the many different design parameters and terrain characteristics can described. By comparison with experiments it has been shown that this description is fairly good approximation of reality.

While the analyses of this chapter are at times complex and algebraically messy, they truly address a very important industrial applications area. Most commercial bulldozers and tracked agricultural machines have flexible tracks and their field behaviour – if one includes track static sinkage, slip sinkage and non-linear under-track soil failure – is very complex. Unfortunately, a complex multi-factorial theory is required to describe such complex real world behaviour.

Finally, and in conclusion, this book has really been an attempt to mathematically-model complex systems. To the extent that the reader considers that this aim has been achieved methodologically and to the extent that the specific models developed are useful, this book will have achieved success.

REFERENCES

1. Wong, J.Y. (1989). *Terramechanics and Off-Road Vehicles.* pp. 105–137. Elsevier.
2. Muro, T. (1989). Stress and Slippage Distributions under Track Belt Running on a Weak Terrain. *Soils and Foundations, Vol. 29, No. 3*, pp. 115–126.
3. Muro, T. (1989). Tractive Performance of a Bulldozer Running on Weak Terrain. *J. of Terramechanics, Vol. 26, No. 3/4*, pp. 249–273.

4. Muro, T. (1995). Trafficability Control System for a Tractor Travelling up and down a Weak Slope Terrain using Initial Track Belt Tension. *Soils and Foundations, Vol. 35, No. 1*, pp. 55–64.

5. Muro, T. (1991). Optimum Track Belt Tension and Height of Application Forces of a Bulldozer Running on Weak Terrain. *J. of Terramechanics, Vol. 28, No. 2/3*, pp. 243–268.

6. Karafiath, L.L. & Nowatzki, E.A. (1978). *Soil Mechanics for Off-Road Vehicle Engineering.* pp. 429–462. Trans Tech Publications.

7. Rowland, D. (1972). Tracked Vehicle Ground Pressure and its Effect on Soft Ground Performance. *Proc. 4th Int. Conf. ISTVS, Stockholm, Sweden, Vol. 1*, pp. 353–384.

8. Cleare G.V. (1971). Some Factors which Influence the Choice and Design of High-Speed Track Layers. *J. of Terramechanics, Vol. 8, No. 2.*

9. Fujii, H., Sawada, T. & Watanabe, T. (1984). Stresses in situ Generating by Bulldozers. *Proc. 8th Int. Conf. ISTVS, Cambridge, England, Vol. 1*, pp. 259–276.

10. Sofiyan, A.P. & Maximenko, Ye.I. (1965). The Distribution of Pressure under a Tracklaying Vehicle. *J. of Terramechanics, Vol. 2, No. 3*, p. 11.

11. Yong, R.N. & Fattah, E.A. (1975). Influence of Contact Characteristics on Energy Transfer and Wheel Performance on Soft Soil. *Proc. 5th Int. Conf. ISTVS, Detroit, U.S.A., Vol. 2* pp. 291–310.

12. Sugiyama, N. (1976). Traffic Performance of a Crawler. *Construction Machinery and Soil.* pp. 56–63. Japan Industrial Press. (In Japanese).

13. Hata, S. (1987). *Theory of Construction Machinery.* pp. 91–93. Kashima Press. (In Japanese).

14. Bekker, M.G. (1956). *Theory of Land Locomotion.* pp. 186–244. The University of Michigan Press.

15. Bekker, M.G. (1960). *Theory of Land Locomotion.* pp. 101–112. The University of Michigan Press.

16. Wills, B.M.D. (1963). The Measurement of Soil Shear Strength and Deformation Moduli and a Comparison of the Actual and Theoretical Performance of a Family of Rigid Tracks. *J. of Agricultural Engineering Research, Vol. 8, No. 2.*

17. Torii, T. (1976). Trafficability of Construction Machinery. *Construction Machinery and Soil.* pp. 47–55. Japan Industrial Press. (In Japanese).

18. Wong, J.Y. (1989). *Terramechanics and Off-Road Vehicles.* pp. 121–137. Elsevier.

19. Ito, G., Maeda, T. & Ohta, H. (1983). Contact Pressure Distribution of a Tracked Vehicle. *Proc. of Symp. on Construction Machinery and Works.* pp. 17–20.

20. Bekker, M.G. (1969). *Introduction to Terrain-Vehicle Systems.* pp. 482–491. The University of Michigan Press.

21. Yong, R.N., Elmamlouk, H. & Della-moretta, L. (1980). Evaluation and Prediction of Energy Losses in Track-Terrain Interaction, *J. of Terramechanics, Vol. 17, No. 2*, pp. 79–100.

22. Yong, R.N., Fattah, E.A. & Skiadas, N. (1984). *Vehicle Traction Mechanics.* pp. 195–255. Elsevier.

23. Garber, M. & Wong, J.Y. (1981). Prediction of Ground Pressure Distribution under Tracked Vehicles, *J. Terramechanics, Vol. 18, No. 1*, pp. 1–23.

24. Wong, J.Y. (1986). Computer aided Analysis of the Effects of Design Parameters on the Performance of Tracked Vehicles. *J. Terramechanics, Vol. 23, No. 2*, pp. 95–124.

25. Oida, A. (1976). Analysis of Tractive Performance of Track-Laying Tractors, *J. of Agricultural Machinery, Vol. 38, No. 1*, pp. 25–40. (In Japanese).

26. Muro, T. (1990). Automated Tension Control System of Track Belt for Bulldozing Operation. *Proc. of the 7th Int. Symp. on Automation and Robotics in Construction, Bristol Polytechnic, Bristol, England.* pp. 415–422.

27. Muro, T. (1991). Tension Control System of Track belt of a Tractor Carrying Down Weak Slope at Braking State. *Proc. 2nd Symp. on Construction Robotics in Japan.* pp. 41–50, JSCE et al. (In Japanese).

28. Muro, T. (1991). Initial Track Belt Tension Affecting on the Tractive Performance of a Bulldozer Running on Weak Terrain. *Terramechanics, Vol. 11*, pp. 15–20. The Japanese Society for Terramechanics. (In Japanese).

EXERCISES

(1) Show that the effective input energy per second E_1 of a flexible tracked vehicle during driving action can be expressed as the sum of the compaction energy E_2, the slippage energy E_3, the effective driving force energy E_4, and the potential energy E_5.

(2) Also, show that the effective input energy per second E_1 of a flexible tracked vehicle during braking action can be expressed as the sum of the compaction energy E_2, the slippage energy E_3, the effective braking force energy E_4, and the potential energy E_5.

(3) Compare the relationships between the amount of sinkage of the front-idler and the rear sprocket and slip ratio which are shown in Figure 6.13 for the driving state and in Figure 6.20 for the braking state. Explain the fact that, for the same absolute value of slip ratio and skid i.e. $i_d = |i_b|$, the amount of sinkage of the rear sprocket during driving action is always larger than that which occurs during braking action.

(4) The distribution of the normal contact pressure $p_i'(X)$ on a flexible tracked vehicle is shown in Figure 6.17(a) for the driving state and in Figure 6.23(a) for the braking state. During driving action, the amplitude of the wavy distribution tends to increase toward the forward part of the track belt, while, during braking action, the amplitude tends to increase toward the rear part. Discuss the reasons for these phenomena.

(5) As mentioned in Section 6.6.1 (2), a bulldozer operating on a soft silty loam terrain can be affected to a very large extent by the size of the vehicle. Calculate the maximum drawbar pull T_{4max} for a bulldozer of weights $W = 12.5 \, \text{kN}$ and $200 \, \text{kN}$ respectively through use of the terrain-track system constants given in Table 6.3.

(6) The effect of the initial track belt tension H_0 on the tractive power efficiency E_d of a bulldozer operating on a soft silty loam terrain has been presented in Section 6.6.1 (3). The dimensions of a particular bulldozer are given in Table 6.5. Calculate the ratio of variation of the tractive power efficiency E_d for $H_0 = 0 \, \text{kN}$ to $50 \, \text{kN}$, for the case where the amount of eccentricity of the center of gravity of the vehicle is $e = 0.00$.

(7) An analytical study of the tractive performances of a small bulldozer operating on a decomposed granite soil has been given in Section 6.6.2 (1). The dimensions of the bulldozer are shown in Table 6.6. Describe in detail the tractive performance of the bulldozer when it is operating under maximum effective tractive effort energy.

(8) The analytical braking performance of a small bulldozer operating on a decomposed granite soil was modeled in Section 6.6.2 (2). The dimensions of the bulldozer are shown in Table 6.6. Describe in detail its braking performance when it is operating under maximum input energy.

(9) An analytical study of the tractive performance of an over snow tracked vehicle of weight $40 \, \text{kN}$ operating on a shallow snow covered terrain of depth $20 \, \text{cm}$ has been given in Section 6.6.3 (1). Describe in detail the tractive performance of the over snow vehicle when it is operating under the condition of maximum drawbar pull energy.

(10) Some analytical results of the braking performances of an over snow tracked vehicle operating on a snow covered terrain have been presented in Section 6.6.3 (2). The vehicle dimensions are as given in Table 6.7. Describe in detail the braking performance of the over snow vehicle when it is operating under conditions of maximum input energy.

Index